Springer Theses

Recognizing Outstanding Ph.D. Research

Aims and Scope

The series "Springer Theses" brings together a selection of the very best Ph.D. theses from around the world and across the physical sciences. Nominated and endorsed by two recognized specialists, each published volume has been selected for its scientific excellence and the high impact of its contents for the pertinent field of research. For greater accessibility to non-specialists, the published versions include an extended introduction, as well as a foreword by the student's supervisor explaining the special relevance of the work for the field. As a whole, the series will provide a valuable resource both for newcomers to the research fields described, and for other scientists seeking detailed background information on special questions. Finally, it provides an accredited documentation of the valuable contributions made by today's younger generation of scientists.

Theses may be nominated for publication in this series by heads of department at internationally leading universities or institutes and should fulfill all of the following criteria

- They must be written in good English.
- The topic should fall within the confines of Chemistry, Physics, Earth Sciences, Engineering and related interdisciplinary fields such as Materials, Nanoscience, Chemical Engineering, Complex Systems and Biophysics.
- The work reported in the thesis must represent a significant scientific advance.
- If the thesis includes previously published material, permission to reproduce this must be gained from the respective copyright holder (a maximum 30% of the thesis should be a verbatim reproduction from the author's previous publications).
- They must have been examined and passed during the 12 months prior to nomination.
- Each thesis should include a foreword by the supervisor outlining the significance of its content.
- The theses should have a clearly defined structure including an introduction accessible to new PhD students and scientists not expert in the relevant field.

Indexed by zbMATH.

Masataka Koide

Non-invertible Symmetry in 4-Dimensional $\mathbb{Z}_2$ Lattice Gauge Theory

Doctoral Theses accepted by
Osaka University, Toyonaka, Osaka, Japan

Springer

Author
Dr. Masataka Koide
Osaka University
Toyonaka, Osaka, Japan

Supervisor
Prof. Tetsuya Onogi
Osaka University
Toyonaka, Osaka, Japan

ISSN 2190-5053 ISSN 2190-5061 (electronic)
Springer Theses
ISBN 978-981-96-2271-9 ISBN 978-981-96-2272-6 (eBook)
https://doi.org/10.1007/978-981-96-2272-6

This Springer imprint is published by the registered company Springer Nature Singapore Pte Ltd.
The registered company address is: 152 Beach Road, #21-01/04 Gateway East, Singapore 189721, Singapore

Supervisor's Foreword

It is my greatest pleasure to introduce the work of Dr. Masataka Koide for publication in the Springer Thesis. Dr. Koide's research focuses on the "non-invertible symmetry," a topic that has undergone significant development in recent years.

Symmetries hold pivotal significance in field theory, both conceptually and practically. Recent advancements have broadened the concept of symmetry, characterizing it as a form of topological defect known as "generalized symmetry." Within this framework, "non-invertible symmetries" emerge as those lacking a group structure.

Until now, research on non-invertible symmetries has primarily centered on two-dimensional systems, yielding numerous intriguing findings. Yet, the exploration of non-invertible symmetry in four-dimensional field theory, which is of interest in elementary particle theory, has remained relatively sparse. Notably, an unresolved question pertains to the existence of non-invertible symmetry, as expressed by three-dimensional topological defects in four dimensions.

Dr. Koide's research represents a pioneering effort, presenting the first example of a three-dimensional topological defect expressing non-invertible symmetry in four-dimensional field theory. In resolving the aforementioned open question, Dr. Koide further revealed the algebraic structure of this symmetry. Such advancements mark a significant stride in the study of non-invertible symmetry in four dimensions. Applying the method similar to that developed by Aasen, Mong and Fendley for the two-dimensional lattice Ising model, Dr. Koide's research extends to the four-dimensional $\mathbb{Z}_2$ lattice gauge theory. This research has reverberated across the global research community, fostering notable progress in the study of non-invertible symmetries beyond two dimensions.

Furthermore, Dr. Koide's research extends to the practical application of the newfound non-invertible symmetry in four dimensions. The partition function on the hemisphere is called the g-function, and it is one of the important quantities that indicate the direction of the flow of the renormalization group. It was demonstrated that the ratio of g-functions can be determined exactly using non-invertible symmetry. This eliminates the potential scenarios of renormalization group flow, offering insights with implications for future studies.

Dr. Koide's groundbreaking studies have not only unveiled the prospect of non-invertible symmetries in higher dimensions but have also exerted a profound influence on researchers worldwide.

Osaka, Japan
December 2024

Tetsuya Onogi

Preface

Symmetry is one of the crucial tools in the non-perturbative analysis in quantum field theory. In the last decade, generalized symmetry, which considers topological defects as symmetries, has been actively studied. One of the generalizations of the symmetry is the non-invertible symmetry described by topological defects without group structure. Non-invertible symmetry has been actively studied in two dimensions, but the theory in higher dimensions has been less well understood than in two dimensions due to a lack of concrete examples. In particular, 3-dimensional non-invertible symmetry defects had not been found in 4-dimensional theories. We consider a 4-dimensional $\mathbb{Z}_2$ pure lattice gauge theory to construct 3-dimensional non-invertible defects. The theory has 1-form $\mathbb{Z}_2$ global symmetry and Kramers-Wannier-Wegner (KWW) duality. We construct 3-dimensional topological defects on the 4-dimensional lattice gauge theory by applying the Aasen-Mong-Fendley approach, which constructs a non-invertible topological defect from the duality in the 2-dimensional Ising model. We show that these defects are non-invertible. We also construct 2-dimensional topological defects corresponding to the 1-form $\mathbb{Z}_2$ global symmetry and a topological junction connecting the non-invertible symmetry and the 1-formsymmetry. We determined the relationship between defects with different topologies and the expectation value of the non-invertible symmetry defects in several configurations.

We also consider theories with boundaries and in particular define three types of boundary conditions in our setup. We considered the conditions for the non-invertible symmetry to have edges on those boundaries. We then find edges of non-invertible defects that can be moved topologically on the boundary. The boundary conditions change on both sides of the edge of the topological defect. We obtained ratios between partition functions with different boundary conditions by topologically deforming the non-invertible defects ending on the boundary. This result provides a restriction on the renormalization group flow.

Toyonaka, Japan Masataka Koide

Parts of this thesis have been published in the following journal articles:

- Koide, Masataka and Nagoya, Yuta and Yamaguchi, Satoshi. Non-invertible topological defects in 4-dimensional $\mathbb{Z}_2$ pure lattice gauge theory. *PTEP*, 2022(1):013B03, 2022
- Masataka Koide, Yuta Nagoya and Satoshi Yamaguchi. Non-invertible symmetries and boundaries in four dimensions. *Phys. Rev. D*, 108(6):065009, 2023.

Acknowledgments

First and foremost, I would like to express my sincere gratitude to Prof. Satochi Yamaguchi for his tremendous support and guidance. He has supported my research activities and provided me with a great deal of knowledge and deep insights. Without his guidance, I have not been able to complete this thesis.

I am also deeply grateful to my collaborator, Yuta Nagoya. I had many discussions with him and learned many things through them. I also greatly appreciate the helpful comments on my doctoral dissertation and presentation by Prof. Tetsuya Onogi, Prof. Tatsuma Nishioka, Prof. Satoshi Yukawa and Prof. Hidenori Fukaya.

I am also grateful to the faculty members of Particle Physics Theory group of Osaka University, Prof. Shinya Kanemura, Prof. Ryosuke Sato, Prof. Minoru Tanaka, Prof. Norihiro Iizuka, Prof. Kei Yagyu, Prof. Yutaka Hosotani and the secretaries, Ms. Kazumi Asano and Ms. Akiko Takao. I would like to thank to all members of my group.

I would also like to express my deep gratitude to Akiyuki Tokuno, Smilin Prince Nelson and Arun Kumar Anbalagan of Springer Nature for their support in the publication of this work. This thesis could only have been published with their help.

Finally, I would like to express my deepest gratitude to my family for the warm support and encouragements.

Contents

Chapter 1
Introduction

Symmetry is an essential tool in the non-perturbative analysis of quantum field theory and plays a central role in understanding phenomena such as anomalies and spontaneous symmetry breaking. In particular, non-perturbative analysis based on the 't Hooft anomaly, an obstruction in the gauging of the global symmetry, has been actively studied in recent years.

In the past decade, the notion of topological defects as generalized symmetries [1] has been actively studied. A topological defect is defined on a nonlocal surface and refers to an operator that can change its shape topologically without affecting its expectation value. According to generalized symmetry, conventional ordinary symmetry is expressed as an invertible topological defect defined on $(d-1)$-dimensions when the dimension of spacetime is d. Those defects then act on the charged operators according to their symmetry transformations. Invertibility implies the existence of a topological defect that produces the inverse transformation. Furthermore, in higher-form symmetry, topological defects defined on a surface of dimension lower than $(d-1)$ are considered as symmetries. In non-invertible symmetry, topological defects without invertible transformations are also treated as symmetries. Other examples of generalized symmetries are higher group symmetries that are algebraic by combining higher-form symmetries of different orders, and subsystem symmetries such that they are topological only in a particular direction in spacetime. These notions of generalized symmetries, coupled with analyses provide a deeper understanding of field theory.

This paper focuses specifically on non-invertible symmetry. Although non-invertible symmetry is a recently named concept based on the notion of generalized symmetry, the study of non-invertible topological operators has existed in the background for some time. In particular, there had been extensive research on the 2-dimensional theory [2–23]. However, at the time we started our research, there were few concrete examples and applications in higher dimensions [24–33], and in particular, non-invertible symmetry defined on a 3-dimensional surface in 4-dimensional spacetime was considered an open question, whether it even existed.

M. Koide, *Non-invertible Symmetry in 4-Dimensional Z2 Lattice Gauge Theory*,
Springer Theses, https://doi.org/10.1007/978-981-96-2272-6_1

We focused on duality as the key to finding non-invertible symmetries defined on 3-dimensional surfaces in 4 dimensions. A duality means that there is more than one theory describing one phenomenon. When there is a duality relationship with oneself, such duality is called self duality. When there is duality, we can obtain physical quantities in two different calculations according to each theory, and those calculations give the same quantity. Therefore, we expect the existence of topological defects corresponding to duality transformations that change these descriptions. And if the duality is self-dual, then the topological defect can be regarded as a topological defect defined on one theory. Moreover, since this defect is placed at the boundary between dual theories, it is defined on a 3-dimensional surface in 4-dimensional spacetime. In fact, in the 2-dimensional Ising model, a non-invertible topological defect was constructed by Aasen-Mong-Fendley (AMF) [34, 35] based on Kramers-Wannier (KW) duality [36, 37]. Therefore, we have turned our attention to the 4-dimensional $\mathbb{Z}_2$ pure lattice gauge theory [38]. In this model, there exists Kramers-Wannier-Wegner (KWW) duality [39], which is generalization of the KW duality. We applied the AMF approach [34] to this theory and actually constructed a topological defect. And we proved that the defects are non-invertible defects [40]. This KWW duality defect is the first concrete example of a 3-dimensional non-invertible defect in 4-dimensional spacetime.

KW duality and its generalization, KWW duality, are dualities associated with gauging global symmetry. And now those duality defects are understood as one of the topological defects that arise when gauging a part of spacetime on a self-dual theory [41, 42]. Another way to produce non-invertible symmetries has been found to gauge higher-form symmetries only on a surface, rather than on the whole of spacetime [43]. These methods have led to the discovery of many non-invertible symmetries of higher dimensions, and their applications as well as methods for finding symmetries are now being actively discussed [41–57]. Lecture notes and reviews on generalized symmetry include the following [58–67].

In this thesis, we describe our work [68] as one of the applications of non-invertible symmetry in higher dimensions. We restrict the renormalization group flow by using duality defects. It is known that in 2-dimensional conformal field theory (CFT), the conformal anomaly decreases monotonically along the renormalization group flow. The property is called the c-theorem [69]. A similar statement is known as the a-theorem in four dimensions [70–73]. A similar statement is also expected in conformal field theories with boundary [74–77]; The hemisphere partition function with a given conformal boundary condition is monotonically decreasing in two and three dimensions and monotonically increasing in four dimensions along the renormalization group flow. This statement is proved in two [78, 79], three [80], and four dimensions [81].

Since duality defects are three-dimensional defects, their edge is the interface between two boundary conditions. We construct an edge that can move topologically on the boundary. We then show that the duality defect can be replaced by a c-number if no operator is inserted in the S^3 region bounded by the spacetime boundary and the bulk duality defect with a topology of D^2, respectively. The boundary condition after the transformation is the boundary condition that originally extended outside of S^3. With this we obtained the ratio between the hemisphere partition functions of the 4-dimensional $\mathbb{Z}_2$ lattice gauge theory with different boundary conditions. From this result and the g-theorem, we restrict the renormalization group flow.

In Chap. 2 we begin with a discussion of ordinary symmetry and introduce the concept of generalized symmetry. In Chap. 3, we consider the 2-dimensional Ising model and confirm that invertible topological defects can be constructed by the AMF approach. In Chap. 4, we construct non-invertible topological defects based on KWW duality in 4-dimensional $\mathbb{Z}_2$ pure lattice gauge theory. We also construct topological defects corresponding to 1-form symmetry and study their algebraic structure. In Chap. 5, we consider topological connections of KWW duality defects to the boundary of spacetime. Using KWW duality defects with edges at the boundary, we compute the ratio between hemisphere partition functions for different boundary conditions. Using the results, we impose restrictions on the renormalization group flow.

This thesis is based on the following works [40, 68].

References

1. Gaiotto, D., et al.: Generalized global symmetries. JHEP **02**, 172 (2015). https://doi.org/10.1007/JHEP02(2015)172. arXiv:1412.5148 [hep-th]
2. Verlinde, E.: Fusion rules and modular transformations in 2D conformal field theory. Nucl. Phys. B **300**, 360–376 (1988). ISSN: 0550-3213
3. Moore, G.W., Seiberg, N.: Classical and quantum conformal field theory. Commun. Math. Phys. **123**, 177 (1989). https://doi.org/10.1007/BF01238857
4. Moore, G.W., Seiberg, N.: Taming the conformal zoo. Phys. Lett. B **220**, 422–430 (1989). https://doi.org/10.1016/0370-2693(89)90897-6
5. Petkova, V.B., Zuber, J.B.: Generalized twisted partition functions. Phys. Lett. B **504**, 157–164 (2001). https://doi.org/10.1016/S0370-2693(01)00276-3. arXiv:hep-th/0011021
6. Fuchs, J., Runkel, I., Schweigert, C.: TFT construction of RCFT correlators 1. Partition functions. Nucl. Phys. B **646**, 353–497 (2002). https://doi.org/10.1016/S0550-3213(02)00744-7. arXiv:hep-th/0204148
7. Frohlich, J., et al.: Kramers-Wannier duality from conformal defects. Phys. Rev. Lett. **93**, 070601 (2004). https://doi.org/10.1103/PhysRevLett.93.070601. arXiv:cond-mat/0404051
8. Carqueville, N., Runkel, I.: Orbifold completion of defect bicategories. Quantum Topol. **7**, 203 (2016). https://doi.org/10.4171/QT/76. arXiv:1210.6363 [math.QA]
9. Brunner, I., Carqueville, N., Plencner, D.: A quick guide to defect orbifolds. In: Donagi, R., et al. (eds.) Proceedings of Symposia in Pure Mathematics, vol. 88, pp. 231–242 (2014). https://doi.org/10.1090/pspum/088/01456. arXiv:1310.0062 [hep-th]
10. Bhardwaj, L., Tachikawa, Y.: On finite symmetries and their gauging in two dimensions. JHEP **03**, 189 (2018). https://doi.org/10.1007/JHEP03(2018)189. arXiv:1704.02330 [hep-th]
11. Chang, C.-M., et al.: Topological defect lines and renormalization group flows in two dimensions. JHEP **01**, 026 (2019). https://doi.org/10.1007/JHEP01(2019)026. arXiv:1802.04445 [hep-th]
12. Freed, D.S., Teleman, C.: Topological dualities in the Ising model (2018). arXiv: 1806.00008 [math.AT]
13. Lin, Y.-H., Shao, S.-H.: Duality defect of the monster CFT. J. Phys. A **54**(6), 065201 (2021). https://doi.org/10.1088/1751-8121/abd69e. arXiv:1911.00042 [hep-th]

14. Thorngren, R., Wang, Y.: Fusion category symmetry i: anomaly in-flow and gapped phases (2019). arXiv: 1912.02817 [hep-th]
15. Gaiotto, D., Kulp, J.: Orbifold groupoids. JHEP **02**, 132 (2021). https://doi.org/10.1007/JHEP02(2021)132. arXiv:2008.05960 [hep-th]
16. Komargodski, Z., et al.: Symmetries and strings of adjoint QCD2. JHEP **03**, 103 (2021). https://doi.org/10.1007/JHEP03(2021)103. arXiv:2008.07567 [hep-th]
17. Huang, T.-C., Lin, Y.-H.: Topological field theory with Haagerup symmetry (2021). arXiv: 2102.05664 [hep-th]
18. Inamura, K.: Topological field theories and symmetry protected topological phases with fusion category symmetries. JHEP **05**, 204 (2021). https://doi.org/10.1007/JHEP05(2021)204. arXiv: 2103.15588 [cond-mat.str-el]
19. Thorngren, R., Wang, Y.: Fusion category symmetry II: categoriosities at c = 1 and beyond (2021). arXiv: 2106.12577 [hep-th]
20. Sharpe, E.: Topological operators, noninvertible symmetries and decomposition (2021). arXiv: 2108.13423 [hep-th]
21. Huang, T.-C., Lin, Y.-H., Seifnashri, S.: Construction of two-dimensional topo logical field theories with non-invertible symmetries. JHEP **12**, 028 (2021). https://doi.org/10.1007/JHEP12(2021)028. arXiv: 2110.02958 [hep-th]
22. Inamura, K.: Fermionization of fusion category symmetries in 1+1 dimensions (2022). arXiv: 2206.13159 [cond-mat.str-el]
23. Lin, Y.-H., et al.: Asymptotic density of states in 2d CFTs with non-invertible symmetries. JHEP **03**, 094 (2023). https://doi.org/10.1007/JHEP03(2023)094. arXiv:2208.05495 [hep-th]
24. Kapustin, A., Saulina, N.: Surface operators in 3d topological field theory and 2d rational conformal field theory. In: Sati, H., Schreiber, U. (eds.) (2010). arXiv: 1012.0911 [hep-th]
25. Ji, W., Wen, X.-G.: Categorical symmetry and noninvertible anomaly in symmetry breaking and topological phase transitions. Phys. Rev. Res. **2**(3), 033417 (2020). https://doi.org/10.1103/PhysRevResearch.2.033417. arXiv:1912.13492 [cond-mat.str-el]
26. Kong, L., et al.: Algebraic higher symmetry and categorical symmetry—a holographic and entanglement view of symmetry. Phys. Rev. Res. **2**(4), 043086 (2020). https://doi.org/10.1103/PhysRevResearch.2.043086. arXiv: 2005.14178 [cond-mat.str-el]
27. Rudelius, T., Shao, S.-H.: Topological operators and completeness of spectrum in discrete gauge theories. JHEP **12**, 172 (2020). https://doi.org/10.1007/JHEP12(2020)172. arXiv:2006.10052 [hep-th]
28. Johnson-Freyd, T.: (3+1)D topological orders with only a Z2-charged particle (2020). arXiv: 2011.11165 [math.QA]
29. Nguyen, M., Tanizaki, Y., Unsal, M.: Semi-Abelian gauge theories, non-invertible symmetries, and string tensions beyond N-ality. JHEP **03**, 238 (2021). https://doi.org/10.1007/JHEP03(2021)238. arXiv:2101.02227 [hep-th]
30. Heidenreich, B., et al.: Non-invertible global symmetries and completeness of the spectrum (2021). arXiv: 2104.07036 [hep-th]
31. Nguyen, M., Tanizaki, Y., Unsal, M.: Non-invertible 1-form symmetry and Casimir scaling in 2d Yang-Mills theory (2021). arXiv: 2104.01824 [hep-th]
32. Delmastro, D., Gomis, J., Yu, M.: Infrared phases of 2d QCD (2021). arXiv: 2108.02202 [hep-th]
33. Kong, L., Wen, X.-G., Zheng, H.: One dimensional gapped quantum phases and enriched fusion categories (2021). arXiv: 2108.08835 [cond-mat.str-el]
34. Aasen, D., Mong, R.S.K., Fendley, P.: Topological defects on the lattice I: the Ising model. J. Phys. A **49**(35), 354001 (2016). https://doi.org/10.1088/1751-8113/49/35/354001. arXiv:1601.07185 [cond-mat.stat-mech]
35. Aasen, D., Fendley, P., Mong, R.S.K.: Topological defects on the lattice: dualities and degeneracies (2020). arXiv: 2008.08598 [cond-mat.stat-mech]
36. Kramers, H.A., Wannier, G.H.: Statistics of the two-dimensional ferromagnet. Part I. Phys. Rev. **60**, 252 (1941). https://doi.org/10.1103/PhysRev.60.252

37. Kramers, H.A., Wannier, G.H.: Statistics of the two-dimensional ferromagnet. Part II. Phys. Rev. **60**, 263 (1941). https://doi.org/10.1103/PhysRev.60.263
38. Wilson, K.G.: Confinement of quarks. Phys. Rev. D **10**, 2445–2459 (1974). Taylor, J.C. (ed.). https://doi.org/10.1103/PhysRevD.10.2445.
39. Wegner, F.J.: Duality in generalized Ising models and phase transitions without local order parameters. J. Math. Phys. **12**, 2259–2272 (1971). https://doi.org/10.1063/1.1665530
40. Koide, M., Nagoya, Y., Yamaguchi, S.: Non-invertible topological defects in 4- dimensional Z2 pure lattice gauge theory. PTEP **2022**(1), 013B03 (2022). arXiv: 2109.05992 [hep-th]
41. Choi, Y., et al.: Noninvertible duality defects in 3+1 dimensions. Phys. Rev. D **105**(12), 125016 (2022). https://doi.org/10.1103/PhysRevD.105.125016. arXiv:2111.01139 [hep-th]
42. Kaidi, J., Ohmori, K., Zheng, Y.: Kramers-wannier-like duality defects in (3+1)D gauge theories. Phys. Rev. Lett. **128**(11), 111601 (2022). https://doi.org/10.1103/PhysRevLett.128.111601. arXiv:2111.01141 [hep-th]
43. Roumpedakis, K., Seifnashri, S., Shao, S.-H.: Higher gauging and non-invertible condensation defects (2022). arXiv: 2204.02407 [hep-th]
44. Wang, J., You, Y.-Z.: Gauge enhanced quantum criticality between grand unifications: categorical higher symmetry retraction (2021). arXiv: 2111.10369 [hep-th]
45. Bhardwaj, L., et al.: Non-invertible higher-categorical symmetries. SciPost Phys. **14**(1), 007 (2023). arXiv: 2204.06564 [hep-th]
46. Hayashi, Y., Tanizaki, Y.: Non-invertible self-duality defects of Cardy-Rabinovici model and mixed gravitational anomaly. JHEP **08**, 036 (2022). https://doi.org/10.1007/JHEP08(2022)036. arXiv:2204.07440 [hep-th]
47. Choi, Y., et al.: Non-invertible condensation, duality, and triality defects in 3+1 dimensions (2022). arXiv: 2204.09025 [hep-th]
48. Choi, Y., Lam, H.T., Shao, S.-H.: Noninvertible global symmetries in the standard model. Phys. Rev. Lett. **129**(16), 161601 (2022). https://doi.org/10.1103/PhysRevLett.129.161601. arXiv:2205.05086 [hep-th]
49. Cordova, C., Ohmori, K.: Noninvertible chiral symmetry and exponential hierarchies. Phys. Rev. X **13**(1), 011034 (2023). https://doi.org/10.1103/PhysRevX.13.011034. arXiv:2205.06243 [hep-th]
50. Bhardwaj, L., Schafer-Nameki, S., Wu, J.: Universal non-invertible symmetries. Fortsch. Phys. **70**(11), 2200143 (2022). arXiv: 2208.05973 [hep-th]
51. Bhardwaj, L., Schafer-Nameki, S., Tiwari, A.: Unifying constructions of non invertible symmetries (2022). arXiv: 2212.06159 [hep-th]
52. Kaidi, J., et al.: Symmetry TFTs and anomalies of non-invertible symmetries. JHEP **10**, 053 (2023). arXiv: 2301.07112 [hep-th]
53. van Beest, M., et al.: Monopoles, scattering, and generalized symmetries (2023). arXiv: 2306.07318 [hep-th]
54. Cordova, C., Hsin, P.-S., Zhang, C.: Anomalies of non-invertible symmetries in (3+1)d (2023). arXiv: 2308.11706 [hep-th]
55. Antinucci, A., et al.: Anomalies of non-invertible self-duality symmetries: fractionalization and gauging (2023). arXiv: 2308.11707 [hep-th]
56. Nagoya, Y., Shimamori, S.: Non-invertible duality defect and non-commutative fusion algebra. JHEP **12**, 062 (2023). https://doi.org/10.1007/JHEP12(2023)062. arXiv:2309.05294 [hep-th]
57. Honda, Y., et al.: Lattice realization of the axial $U(1)$ non-invertible symmetry (2024). arXiv: 2401.01331 [hep-lat]
58. Gomes, P.R.S.: An introduction to higher-form symmetries. SciPost Phys. Lect. Notes **74**, 1 (2023). arXiv: 2303.01817 [hep-th]
59. Bhardwaj, L., et al.: Lectures on generalized symmetries. Phys. Rept. **1051**, 1–87 (2024). arXiv: 2307.07547 [hep-th]
60. Luo, R., Wang, Q.-R., Wang, Y.-N.: Lecture notes on generalized symmetries and applications (2023). arXiv: 2307.09215 [hep-th]
61. Cordova, C., et al.: Snowmass white paper: generalized symmetries in quantum field theory and beyond. In: Snowmass 2021 (2022). arXiv: 2205.09545 [hep-th]

62. Schafer-Nameki, S.: ICTP lectures on (non-)invertible generalized symmetries (2023). arXiv: 2305.18296 [hep-th]
63. Brennan, T.D., Hong, S.: Introduction to generalized global symmetries in QFT and particle physics (2023). arXiv: 2306.00912 [hep-ph]
64. Shao, S.-H.: What's done cannot be undone: TASI lectures on non-invertible symmetry (2023). arXiv: 2308.00747 [hep-th]
65. Yamaguchi, S.: Kramers-wannier duality in 2-dimensional Ising model (in Japanese). http://www-het.phys.sci.osaka-u.ac.jp/
66. Yamaguchi, S.: Non-invertible symmetry on the lattice. http://www-het.phys.sci.osaka-u.ac.jp/
67. Yamaguchi, S.: About generalized symmetry (in Japanese). http://www-het.phys.sci.osaka-u.ac.jp/
68. Koide, M., Nagoya, Y., Yamaguchi, S.: Noninvertible symmetries and boundaries in four dimensions. Phys. Rev. D **108**(6), 065009 (2023). arXiv: 2304.01550 [hep-th]
69. Zamolodchikov, A.B.: Irreversibility of the Flux of the renormalization group in a 2D field theory. JETP Lett. **43**, 730–732 (1986)
70. Cardy, J.L.: Is there a c theorem in four-dimensions? Phys. Lett. B **215**, 749–752 (1988). https://doi.org/10.1016/0370-2693(88)90054-8
71. Osborn, H.: Derivation of a four-dimensionalâŁ¯c-theorem. Phys. Lett. B **222**, 97–102 (1989). https://doi.org/10.1016/0370-2693(89)90729-6
72. Komargodski, Z., Schwimmer, A.: On renormalization group flows in four dimensions. JHEP **12**, 099 (2011). https://doi.org/10.1007/JHEP12(2011)099. arXiv:1107.3987 [hep-th]
73. Casini, H., Landea, I.S., Torroba, G.: Entropic g theorem in general spacetime dimensions. Phys. Rev. Lett. **130**(11), 111603 (2023). https://doi.org/10.1103/PhysRevLett.130.111603. arXiv:2212.10575 [hep-th]
74. Affleck, I., Ludwig, A.W.W.: Universal noninteger 'ground state degeneracy' in critical quantum systems. Phys. Rev. Lett. **67**, 161–164 (1991). https://doi.org/10.1103/PhysRevLett.67.161
75. Nozaki, M., Takayanagi, T., Ugajin, T.: Central charges for BCFTs and holography. JHEP **06**, 066 (2012). https://doi.org/10.1007/JHEP06(2012)066. arXiv:1205.1573 [hep-th]
76. Gaiotto, D.: Boundary F-maximization (2014). arXiv: 1403.8052 [hep-th]
77. Kobayashi, N., et al.: Towards a C-theorem in defect CFT. JHEP **01**, 039 (2019). https://doi.org/10.1007/JHEP01(2019)039. arXiv:1810.06995 [hep-th]
78. Friedan, D., Konechny, A.: On the boundary entropy of one-dimensional quantum systems at low temperature. Phys. Rev. Lett. **93**, 030402 (2004). https://doi.org/10.1103/PhysRevLett.93.030402. arXiv:hep-th/0312197
79. Casini, H., Landea, I.S., Torroba, G.: The g-theorem and quantum information theory. JHEP **10**, 140 (2016). https://doi.org/10.1007/JHEP10(2016)140. arXiv:1607.00390 [hep-th]
80. Casini, H., Landea, I.S., Torroba, G.: Irreversibility in quantum field theories with boundaries. JHEP **04**, 166 (2019). https://doi.org/10.1007/JHEP04(2019)166. arXiv:1812.08183 [hep-th]
81. Casini, H., Landea, I.S., Torroba, G.: Irreversibility, QNEC, and defects (2023). arXiv: 2303.16935 [hep-th]

Chapter 2
Symmetry and Topological Defect

The goal of this chapter is to rewrite and extend the notion of symmetry in terms of topological defects. In particular, we investigate higher-form symmetries as a generalization of symmetries.

2.1 Ordinary Symmetry

In this section, we discuss the usual symmetries. First, we derive the Ward-Takahashi identity (WT identity) of the continuous symmetry. Then we check the local action of the symmetry. Next, we introduce topological defects called symmetry defects in order to apply the local action to more general symmetries, including discrete symmetries. We also confirm that these symmetry defects can be understood as the configuration of flat background gauge fields.

2.1.1 Review of Ordinary Continuous Symmetry

We treat a d-dimensional Euclidean theory. The coordinates are $(x^1, \ldots, x^{d-1}, x^d = \tau)$, where the τ is imaginary time. Let $\phi(x)$ be a field and the Lagrangian density of the theory be $\mathcal{L}(\phi, \partial\phi)$. The action is denoted by

$$S[\phi] = \int d^d x \mathcal{L}(\phi, \partial\phi). \tag{2.1.1}$$

We often state that the theory has symmetry when the action is invariant under field transformations $\phi \to \phi'$.

M. Koide, *Non-invertible Symmetry in 4-Dimensional Z2 Lattice Gauge Theory*,
Springer Theses, https://doi.org/10.1007/978-981-96-2272-6_2

When the theory has continuous symmetry with group G, we can find conserved currents. We consider a continuous symmetry and an infinitesimal global symmetry transformation

$$\phi(x) \to \phi'(x) = \phi(x) + \epsilon\delta\phi(x), \tag{2.1.2}$$

where ϵ is an infinitesimal constant parameter. Then the variation of the Lagrangian $\delta\mathcal{L} \equiv \mathcal{L}(\phi', \partial\phi') - \mathcal{L}(\phi, \partial\phi)$ can be expressed by total derivative because the action is invariant under symmetry transformation,

$$\delta\mathcal{L} = \epsilon\partial_\mu K^\mu(x). \tag{2.1.3}$$

Next, we promote the parameter ϵ of the transformation from constant to coordinate-dependent,

$$\phi(x) \to \phi'(x) = \phi(x) + \epsilon(x)\delta\phi(x). \tag{2.1.4}$$

This transformation is no longer a global transformation, and the action changes under the transformation. the variation of the action is

$$\begin{aligned} \delta S &\equiv S[\phi'] - S[\phi] \\ &= \int d^d x \epsilon(x)\partial_\mu j^\mu(x) \quad \left(j^\mu \equiv -\frac{\partial\mathcal{L}}{\partial\partial_\mu\phi}\delta\phi(x) + K^\mu(x) \right). \end{aligned} \tag{2.1.5}$$

This variation is generally nonzero. However, if the configuration of the field $\phi(x)$ satisfies the equation of motion (EOM), the action is invariant under arbitrary variations. And since the parameter $\epsilon(x)$ is an arbitrary function, we obtain the following conservation law,

$$\partial_\mu j^\mu(x) = 0 \quad \text{(EOM)}. \tag{2.1.6}$$

This conserved current $j^\mu(x)$ is called the Noether current.

We obtain a conserved charge $Q(\Sigma_\tau)$ from this current.

$$Q(\Sigma_\tau) = -i \int_{\Sigma_\tau} d^{d-1}x j^d(x^1, \ldots, x^{d-1}, \tau). \tag{2.1.7}$$

where the Σ_τ is the time-constant surface at time τ. We have defined the dth component of the euclidean current $j^d = ij^0$, where the j^0 is the zeroth component of the Lorentzian current. this definition (2.1.7) consistent with the usual definition in Lorentzian formalism. $Q(\Sigma_\tau)$ is conserved under time translation,

$$\frac{d}{d\tau}Q(\Sigma_\tau) = 0. \tag{2.1.8}$$

We can use this current to consider local actions of the symmetry on the field. To confirm this, let us derive the Ward-Takahashi identity (WT-identity) in path-integral formalism. Let Z be the partition function of the theory and define the correlation function of a generic product of fields $\cdots$ as

$$\langle\cdots\rangle \equiv \frac{1}{Z}\int \mathcal{D}\phi \cdots e^{-S[\phi]}. \tag{2.1.9}$$

We consider a n-point function of the field ϕ and transform it as follows.

$$\begin{aligned}\langle\phi(x_1)\cdots\phi(x_n)\rangle &= \frac{1}{Z}\int \mathcal{D}\phi\phi(x_1)\cdots\phi(x_n)e^{-S[\phi]} \\ &= \frac{1}{Z}\int \mathcal{D}\phi'\phi'(x_1)\cdots\phi'(x_n)e^{-S[\phi']} \\ &= \frac{1}{Z}\int \mathcal{D}\phi\left\{\prod_{i=1}^{n}((\,)\,\phi(x_i)+\epsilon\delta\phi(x_i))\right\}(1-\delta S)e^{-S[\phi]}\end{aligned} \tag{2.1.10}$$

In the second equality, the variables are redefined from the field $\phi(x)$ to the $\phi'(x) = \phi(x) + \epsilon(x)\delta\phi(x)$. The third equality, we assumes that the measure is invariant $\mathcal{D}\phi = \mathcal{D}\phi'$. Using Eq. (2.1.5), we obtain the following equation.

$$\int d^d y\epsilon(y)\left\{\langle\partial_\mu j^\mu(y)\phi(x_1)\cdots\phi(x_n)\rangle - \sum_{i=1}^{n}\delta^d(x_i-y)\,\langle\phi(x_1)\cdots\delta\phi(x_i)\cdots\phi(x_n)\rangle\right\} = 0. \tag{2.1.11}$$

Since $\epsilon(x)$ is an arbitrary function, we can conclude that the contents of braces are zero. Thus we obtain the WT identity.

$$\langle\partial_\mu j^\mu(y)\phi(x_1)\cdots\phi(x_n)\rangle = \sum_{i=1}^{n}\delta^d(x_i-y)\,\langle\phi(x_1)\cdots\delta\phi(x_i)\cdots\phi(x_n)\rangle \tag{2.1.12}$$

When $y \neq x_1, \ldots, x_n$, the right-hand side of the above equation is zero, so the conservation law $\partial_\mu j^\mu = 0$ holds in any correlation function. When $y = x_i$ $(i = 1, \ldots, N)$, the right-hand side has a contact term proportional to $\delta\phi(x_i)$. The WT identity indicates how symmetry transformations act locally.

Let us show how the symmetry transformation acts locally in terms of the conserved charge and the unitary operators. First, we redefine conserved charges on curved surface by

$$Q(M) = -i\int_M dS_\mu j^\mu(x). \tag{2.1.13}$$

Given that $M = \Sigma_\tau$, this definition is consistent with the usual definition (2.1.7). This defects are topological. In other words, its value does not change even if the supported surface M is continuously deformed into other surfaces M'. We can prove this using conservation laws.

$$
\begin{aligned}
Q(M) - Q(M') &= -i \int_{M \cup \bar{M}'} dS_\mu j^\mu(x) \\
&= -i \int_D d^d x \partial_\mu j^\mu(x) = 0.
\end{aligned} \tag{2.1.14}
$$

We denote by $\bar{M}$ the inverse oriented surface of the oriented surface M. The D is the d-dimensional space with boundary $\partial D = M \cup \bar{M}'$. We also define the unitary operator $U_g(M)$ on the surface with a appropriate parameter α as follows,

$$
U_g(M) \equiv e^{i\alpha Q(M)}, \quad (g = e^{i\alpha} \in G, G : \text{group}). \tag{2.1.15}
$$

The unitary defects are topological too.

Consider the WT identity with one inserted field. Integrating the WT identity over a region D with boundary $\partial D = M$ where only $\phi(x_1)$ is inserted, we can describe how the symmetry transformation acts locally, in terms of the conserved charge $Q(M)$.

$$
\begin{aligned}
i \langle Q(M)\phi(x_1)\cdots\rangle &= \int_D d^d y \delta^d(x - y) \langle \delta\phi(x_1)\cdots\rangle . \\
&= \langle \delta\phi(x_1)\cdots\rangle .
\end{aligned} \tag{2.1.16}
$$

It is represented as in Fig. 2.1. This shows that conserved charge defects only act on the fields existing inside the defects.

Conserved charges cause infinitesimal transformations. Finite transformations can be described by unitary operators. Let G be the group structure of the symmetry. Then, by a transformation with respect to the element $g \in G$, we assume that the field

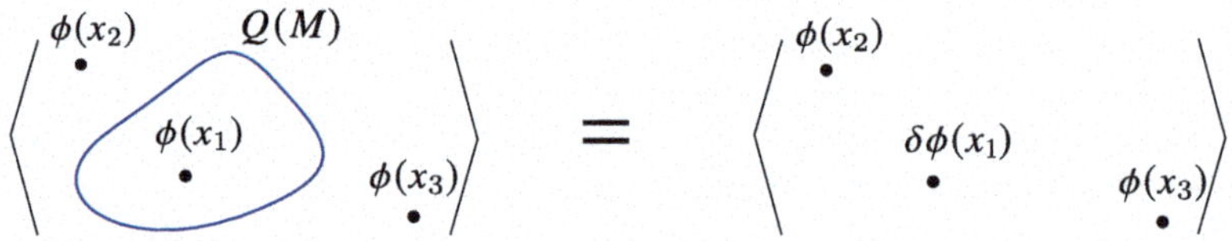

Fig. 2.1 Illustration of the partial action of a conserved charge $Q(M)$. The blue line represents the conserved charge $Q(M)$ defined on a $(d-1)$-dimensional surface M. Only the field $\phi(x_1)$ is inserted in the interior region of the surface M, and the conserved charge $Q(M)$ acts only on $\phi(x_1)$

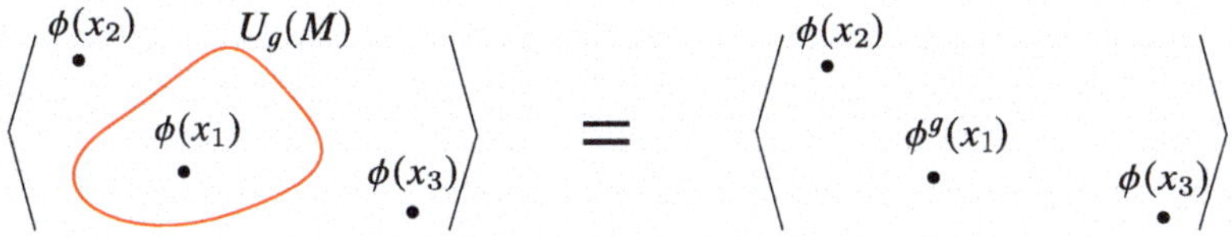

Fig. 2.2 Illustration of the partial action of a unitary operator $U_g(M)$. The red line represents the conserved charge $U_g(M)$ defined on a $(d-1)$-dimensional surface M. Only the field $\phi(x_1)$ is inserted in the interior region of the surface M, and the conserved charge $U_g(M)$ acts only on $\phi(x_1)$

transforms $\phi(x) \to \phi^g(x)$. Finite transformations can be reproduced by repeating infinitesimal transformations, we derive from (2.1.16) that

$$\langle U_g(M)\phi(x_1)\cdots\rangle = \langle \phi^g(x_1)\cdots\rangle . \tag{2.1.17}$$

It is represented as in Fig. 2.2. The unitary defects only act on the fields existing inside the defects too.

As we have seen, when the theory has continuous symmetries, we can construct the current. We obtain a conserved charge $Q(\Sigma_\tau)$ and the unitary operators from this current. Using these, we can compute the local action of the symmetries. However, for discrete symmetries where there is no current, it is not possible to compute the local action by the same method. In the next subsection, we attempt to describe the local action of more general symmetries, including discrete symmetries, by defining unitary operators without via currents.

2.1.2 Topological Defects and Local Action

In this section, we introduce the notion of the topological defects in order to consider their local action even for symmetries without currents, such as discrete symmetries.

First we define a field transformation $\phi \to \phi^{g;D}$ that applies a global transformation only on the d-dimensional subspace D with boundary $\partial D = M$,

$$\phi(x) \to \phi^{g;D}(x) = \begin{cases} \phi^g(x) & (x \in D) \\ \phi(x) & (x \notin D) \end{cases} . \tag{2.1.18}$$

Consider the variation of the action associated with this transformation. This transformation does not transform the field outside of region D, so there is no contribution to the variation from outside of region D. Also, this transformation performs a global transformation on the field inside the region D. However, the internal action is also invariant from symmetry up to surface term. Therefore the variation can be written as an integral over the boundary $\partial D = M$. We denote the variant of the action by

$$\delta_g S(M) \equiv S[\phi^{g;D}] - S[\phi]. \tag{2.1.19}$$

For later use, we derive some properties about the variation. The entire space is written as X. The global transformation for the entire space $\phi^{g;X} = \phi^g$ can be reproduced by simultaneously performing partial global transformations $\phi^{g;D}$ and $\phi^{g;X\backslash D}$. Then, by symmetry $S[\phi^g] = S[\phi]$, we find that the variations of transformations $\phi^{g;D}$ and $\phi^{g;X\backslash D}$ cancel each other. this is expressed as

$$\delta_g S(M) = -\delta_g S(\bar{M}). \tag{2.1.20}$$

Note the orientation of the boundary defined by the region to be transformed. If we repeatedly apply mutually inverse partial global transformations $\phi^{g;D}$ and $\phi^{g^{-1};D}$ to the region D, the field does not change and the action is invariant. From this we obtain

$$\delta_g S(M) = -\delta_{g^{-1}} S(M). \tag{2.1.21}$$

Using this variation, we define the symmetry defect $U_g(M)$ supprted by M as

$$\left\langle U_g(M)\cdots\right\rangle = \frac{1}{Z}\int \mathcal{D}\phi e^{-\delta S_{g^{-1}}(M)}\cdots e^{-S[\phi]}. \tag{2.1.22}$$

By replacing the label ϕ of the integral variable of the correlation function with the partially global transformed field $\phi^{g;D}$, we obtain the correlation function with the symmetry defect inserted. The insertion of the defect is done as follows.

$$\begin{aligned}
\langle\phi(x_1)\cdots\phi(x_n)\rangle &= \frac{1}{Z}\int \mathcal{D}\phi\phi(x_1)\cdots\phi(x_n)e^{-S[\phi]} \\
&= \frac{1}{Z}\int \mathcal{D}\phi^{g;D}\phi^{g;D}(x_1)\cdots\phi^{g;D}(x_n)e^{-S[\phi^{g;D}]} \\
&= \frac{1}{Z}\int \mathcal{D}\phi\phi^{g;D}(x_1)\cdots\phi^{g;D}(x_n)e^{-\delta S(M)}e^{-S[\phi]} \\
&= \left\langle U_{g^{-1}}(M)\phi^{g;D}(x_1)\cdots\phi^{g;D}(x_n)\right\rangle.
\end{aligned} \tag{2.1.23}$$

In the third equality, we assumed that the measure is invariant under label replacement. This is shown in Fig. 2.3. This is WT-identity using the unitary operator (2.1.17) obtained in the previous section. In other words, the symmetry defect is a unitary operator. The symmetry defects are defined from the variation of the action associated with the partial global transformation not via the current. Therefore, this WT-identity is valid for discrete symmetries. By using symmetry defects, we can act locally on discrete symmetries.

Unitary operators are guaranteed to be topological by the conservation laws of the current. Based on the definition of symmetry defects, let us show that symmetry defects are topological. That is, the value does not change under continuous deformations that do not cross the charged operator. We can understand this by the fact that we can freely take the region D in which we perform the partial global transformation

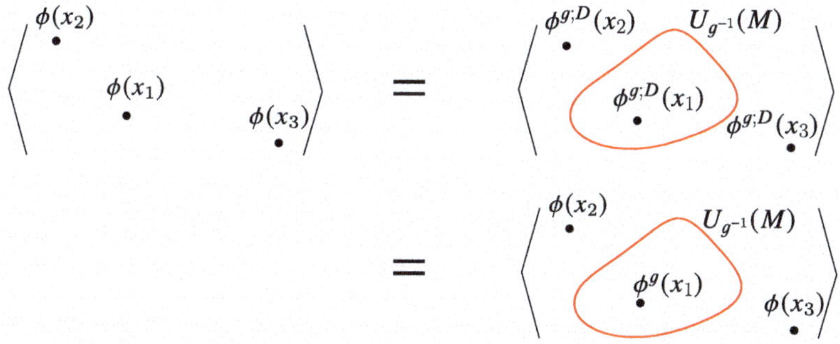

Fig. 2.3 Insertion of a symmetry defect $U_{g^{-1}(M)}$. The red line represents the symmetry defect $U_{g^{-1}(M)}$. The region bounded by the curved surface M is D, and only the field $\phi(x_1)$ is inserted on it. In the first equality, the integral variable is changed from ϕ to $\phi^{g;D}$ and the variation of the action is regarded as the symmetry defect $U_{g^{-1}(M)}$. The second equality uses the fact that $\phi^{g;D}$ is only transformed on D

$\phi^{g^{-1};D}$ used to replace the label. Even if the region D is replaced by another region D' with boudary M' that contains only the same operator inside, the transformation of the operator is the same. Therefore, the defects can be topologically deformed as in Fig. 2.4.

$$\langle U_g(M)\cdots\rangle = \langle U_g(M')\cdots\rangle. \tag{2.1.24}$$

Thus, when a relational expression holds within any correlation function, we omit the bracket and shorthand it as follows,

$$U_g(M) = U_g(M'). \tag{2.1.25}$$

Next, we describe the fusion operation for symmetry defects. Let $D_i (i = 1, 2)$ be a region with boundary $\partial D_i = M_i$. Prepare two symmetry defects $U_{g_1}(M_1)$ and $U_{g_2}(M_2)$. And let D be a region with boundary $\partial D = M_1 \cup \bar{M}_2$. We assume that no field is inserted in this D. Let us now superimpose the symmetry defect $U_{g_1}(M_1)$ on $U_{g_2}(M_2)$ by moving it over D. The resulting defect causes the same transformation

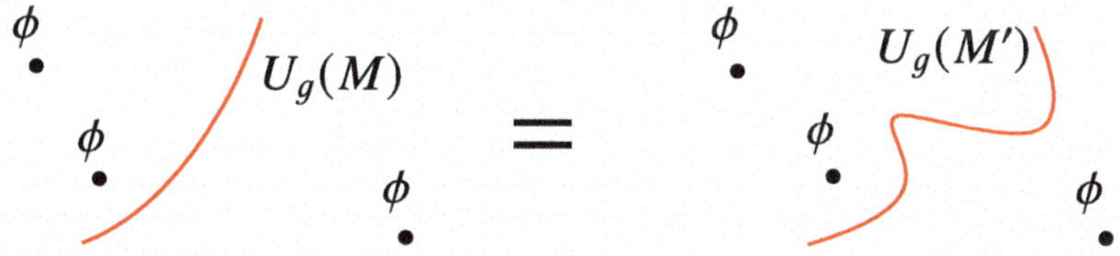

Fig. 2.4 Topological nature of symmetry defects. The red lines represent the unitary defects. The value of the correlation function with symmetry defects does not change when the shape of the symmetry defect is continuously deformed, as long as it does not cross the charged operator

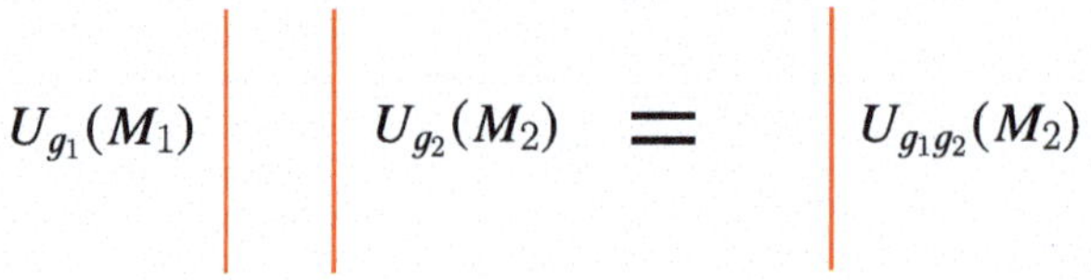

Fig. 2.5 Fusion of symmetry defects. The red lines represent the unitary defects. When two symmetry defects are superimposed on each other, one other symmetry defect is produced

as if we had performed the partial global transformations $\phi^{g_2^{-1};D_2}$ and $\phi^{g_1^{-1};D_2}$ successively, that is $\phi^{(g_1g_2)^{-1};D_2}$, on the operator on the region D_2. Therefore we obtain the following defect relation.

$$U_{g_1}(M_1)U_{g_2}(M_2) = U_{g_1g_2}(M_2). \tag{2.1.26}$$

It is represented as in Fig. 2.5. By this relation we notice that the symmetry defect has a group structure. The identity defect $\mathbf{1}(M)$ is obtained from a identity transformation.

$$\langle \mathbf{1}(M) \cdots \rangle = \frac{1}{Z} \int \mathcal{D}\phi \cdots e^{-S[\phi]}. \tag{2.1.27}$$

From Eqs. (2.1.20), (2.1.21), the variation $\delta_g S(M)$ satisfies $\delta_g S(M) = \delta_{g^{-1}} S(\bar{M})$, we can obtain the inverse element by reversing the orientation of the symmetry defect.

$$U_g(M) = U_{g^{-1}}(\bar{M}). \tag{2.1.28}$$

A illustrated case can be seen in Fig. 2.6.

Using the WT identity, invertibility can be expressed as the fact that a closed invertible topological defect is 1 when there is no internal charged operator as shown in Fig. 2.7.

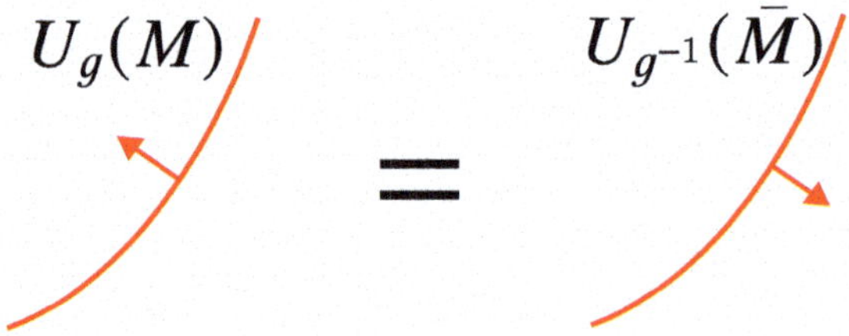

Fig. 2.6 The red lines represent the unitary defects. The red arrows represent the orientation of the surfaces. Reversing the orientation of the surface where the symmetry defect is supported produces an inverse symmetry defect

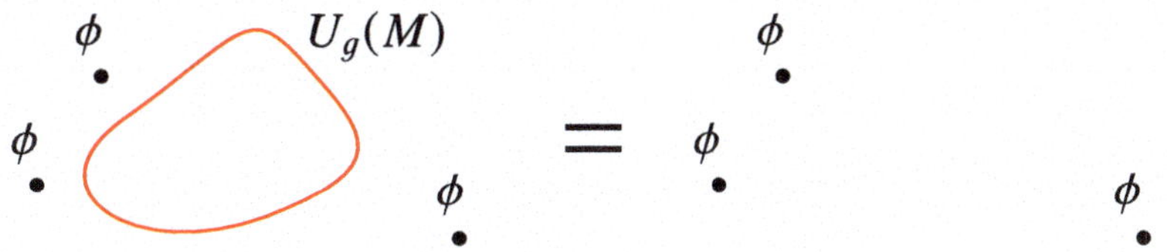

Fig. 2.7 The red lines represent the unitary defects

2.1.3 Symmetry Defects and Background Gauge Field

Next, we introduce the background gauge field B and consider symmetry defects from another perspective. For simplicity, we consider a theory $S[\phi, B]$ with continuous G gauge symmetry. Assume that when $B = 0$, this theory $S[\phi, B = 0]$ becomes the theory $S[\phi]$ with global symmetry G as we have been considering. The partition function is

$$Z[B] = \int \mathcal{D}\phi e^{-S[\phi, B]}. \tag{2.1.29}$$

Since gauge invariance, the partition function $Z[B]$ is invariant even if the background gauge field B is replaced by the gauge transformed field B',

$$Z[B] = Z[B']. \tag{2.1.30}$$

We consider the theory $Z[B = 0]$ and transform the field $\phi \to \phi^{g;D}$ by a gauge transformation. then we denote the gauge-transformed background gauge field by $B^{g;D}$. Since the transformation $\phi \to \phi^{g;D}$ is global inside and outside D, the $B^{g;D}$ is localized on the boundary $M = \partial D$. And the background gauge field $B^{g;D}$ is flat because of $B = 0$. By gauge symmetry, the partition function is invariant.

$$Z[B^{g,D}] = Z[B = 0]. \tag{2.1.31}$$

On the other hand, we can interpret the gauge transformation as the changing the label of the variable from ϕ to $\phi^{g;D}$ in the correlation function. It is inserting the symmetry defect. Therefore we can regard the symmetry defect $U_g(M)$ as the configuration of the localized flat gauge fields $B^{g^{-1};D}$.

Conversely, we can interpret the configuration of the flat gauge field as the configuration of the symmetry defects. Given the flat gauge field, we can gauge transformations so that locally the gauge field is zero. Then we can generally get the configuration of the localized flat gauge field with the junction as in Fig. 2.8. Therefore we can interpret this gauge field as the symmetry defects with the junction. Since we can move the junction of the gauge field by the gauge transformation, the junction of the defects is topological unless crossing the symmetry defects.

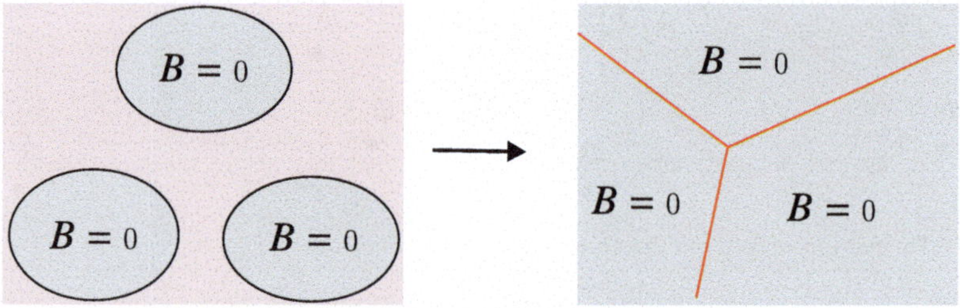

Fig. 2.8 The localization of non-zero gauge field regions by gauge transformation. In red region, the gauge field is non-zero, while in gray regions, the gauge field is zero. In the figure on the right, the non-zero gauge field is localized on the red $(d-1)$-dimensional surface. The intersection of the red lines represents a junction of $(d-1)$-dimensional surfaces

2.1.4 Example: 2-Dimensional Ising Model

In this subsection, we review the construction of symmetry defects in the 2-dimensional Ising model. We also check the relationship between symmetry defects and background gauge fields.

We denote the link with sites i, j at its boundaries by $\langle ij \rangle$ and assign the site variables $U_i = (-1)^{a_i} = \pm 1$ to the site i. Consider following the interaction on the link $l = \langle ij \rangle$,

$$K(-1)^{a_i + a_j}, \tag{2.1.32}$$

where K is a positive constant parameter. We denote the plaquette with sites i, j, k, l on its boundary by $\langle ijkl \rangle$. We define the partition function of the Ising model as

$$Z_{\text{Ising}}(K) = \sum_{\{a\}} \exp\left(K \sum_{l:\text{link}} (-1)^{(a_i + a_j)}\right), \tag{2.1.33}$$

where the sum$\sum_{\{a\}}$ means the summation of the all configuration of the variables a.

The Ising model has $\mathbb{Z}_2$ spin flip symmetry. In fact, considering the spin flip $a_i \to a_i + 1 \pmod 2$, the interaction term is invariant.

We construct the symmetry defects associated with this spin flip symmetry. Let $\tilde{D}$ is a set of the sites and consider the following partial global transformation,

$$a_i \to a_i^{-1;D} = \begin{cases} a_i + 1 \pmod 2 & (i \in \tilde{D}) \\ a_i & (i \notin \tilde{D}) \end{cases}. \tag{2.1.34}$$

We also define the set $\tilde{M}$ as the set of links $\langle ij \rangle$ such that only one of the sites i and j belongs to $\tilde{D}$. By performing this transformation, the interaction of the link $\langle ij \rangle$ changes as follow,

$$K(-1)^{a_i+a_j} \to = \begin{cases} -K(-1)^{a_i+a_j} & (\langle ij\rangle \in \tilde{M}) \\ K(-1)^{a_i+a_j} & (\langle ij\rangle \notin \tilde{M}) \end{cases}. \tag{2.1.35}$$

In other words, only the interactions of the links on the $\tilde{M}$ contribute the variation of the action.

We check where this symmetry defect exist. For this reason, we define the dual lattice putting dual sites at the center of the all plaquette in the original lattice. We define the dual plaquette $\tilde{i}$ as a plaquette in dual lattice with site i in original lattice at its center. We define the dual link $\widetilde{\langle ij\rangle}$ in dual lattice with original plaquette $\langle ij\rangle$ orthogonal to its center. And we define the sets D and M as follows

$$D \equiv \{\tilde{i} \mid i \in \tilde{D}\}, \tag{2.1.36}$$

$$M \equiv \{\widetilde{\langle ij\rangle} \mid \langle ij\rangle \in \tilde{M}\}. \tag{2.1.37}$$

The sets $\tilde{D}$, $\tilde{M}$, D, and M are represented as in Fig. 2.9. The transformation $a^{-1;D}$ act only the site on the 2-dimensional region D. Therefore the $\mathbb{Z}_2$ symmetry defect $\eta(M)$ is defined on the boundary M.

Next we check the correspondence between symmetry defect and background gauge field. We introduce the background gauge field B_{ij} on the link $\langle ij\rangle$. The interaction of the link $\langle ij\rangle$ is modified as follows

$$K(-1)^{a_i+a_j+B_{ij}}. \tag{2.1.38}$$

By the gauge transformation, the site variable and the background gauge field are changed as follow,

$$a_i \to a_i' = a_i + \lambda_i \pmod 2, \tag{2.1.39}$$

$$B_{ij} \to B_{ij}' = B_{ij} + \lambda_i + \lambda_j \pmod 2, \tag{2.1.40}$$

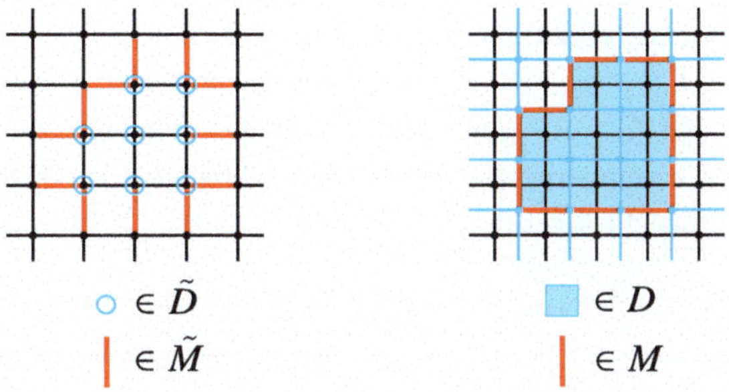

Fig. 2.9 Ising model defined on a 2-dimensional square lattice. Spin variables are assigned on the black dots. In the left figure, the dots with blue rings belong to $\tilde{D}$. The red links belong to $\tilde{M}$ at this time. In the right figure, the blue dots and links represent dual sites and dual links, respectively. The blue plaquettes belong to D. The red dual links belong to M and are part of a symmetry defect

where the $\lambda_i = 0, 1$ is a parameter of the gaugetransformation. Consider a gauge transformation that the parameter $\lambda_i = 1$ only if the site i belong to D. Then the gauge field B_{ij} changes only if the link $\langle ij \rangle$ belong to $\tilde{M}$. When we consider gauge transformation from the background gauge field $B = 0$ to B', the configuration B' is localized on M and flat, that is the B' satisfy the $B_{ij} + B_{jk} + B_{kl} + B_{li} = 0$ (mod2).

2.1.5 *The Notion of Generalized Symmetry*

In this section, by introducing the symmetry defects, we can calculate the local action of the symmetry containing the discrete symmetry. We now summarize the properties of the symmetry defects treated in this section.

1. symmetry defects are topological.
2. symmetry defects are $(d-1)$-dimensional (codimension-1).
3. symmetry defects have group structure.

The codimension-$(d-p)$ means the number of dimensions d of the theory minus the dimension p of the defect. The first property is important and useful because it is related to the conservation law. For example, the topological defects can act on local or nonlocal operators such as the WT identity. So we can consider selection rules associated with the defects. For this reason we consider extending the notion of symmetry by forgetting the second or third property. In other words, we consider the existence of topological defects in the theory as the theory have symmetry. This notion is called generalized symmetry. In particular, symmetries corresponding to topological defects that are not $(d-1)$-dimensional are called higher-form symmetries. If the symmetry defects does not have a group structure, there is generally not an inversed element. So symmetries with topological defects that do not have group structure are called non-invertible symmetries.

2.2 Higher-Form Symmetry

In this section, we consider the p-form symmetry. The p-form symmetry defects are topological and has group structure G. But these defects are codimension-$(p+1)$. The symmetry treated in the previous section is called 0-form symmetry in this perspective.

When the theory has p-form symmetry with group $G^{(p)}$, we can consider p-form symmetry defects $U_g(M)$ on the $d-p-1$ dimensional surface M,

$$U_g(M) \quad (g \in G^{(p)}). \tag{2.2.1}$$

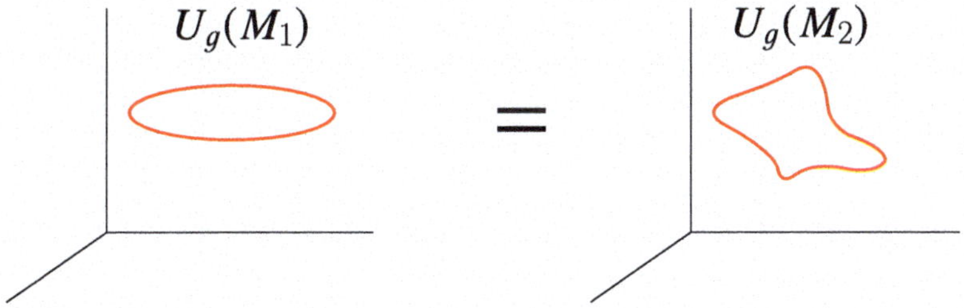

Fig. 2.10 Topological nature of p-form symmetry defects. The red lines represent the p-form unitary defects. The value of the correlation function with p-form symmetry defects does not change when the shape of the symmetry defect is continuously deformed, as long as it does not cross the charged operator

This defects are topological. In other words, Let M_1 and M_2 be two $d-p-1$ dimensional surfaces with the same topology. We can topologically deform the defect $U_g(M_1)$ from M_1 to M_2 without changing the value,

$$U_g(M_1) = U_g(M_2). \tag{2.2.2}$$

This is shown in Fig. 2.10

We show the $p(>0)$-form symmetry group $G^{(p)}$ is Abelian. Place $U_{g_1}(M_1)$ and $U_{g_2}(M_2)$. There exists more than one $d-p$ dimensional manifold with boundary $M_1 \cup \bar{M}_2$. Therefore by the topological deforming, we can exchange the positions of $U_{g_1}(M_1)$ and $U_{g_2}(M_2)$ without crossing each other. So the $p(>0)$-form symmetry group $G^{(p)}$ is Abelian.

As with 0-form symmetry, we can fuse the two p-form symmetry defects as follows,

$$U_{g_1}(M_1)U_{g_2}(M_2) = U_{g_1 g_2}(M_2). \tag{2.2.3}$$

By this fusion operation, the p-form symmetry defects have the group structure $G^{(p)}$.

Next we consider the action of the p-form symmetry defects on the charged operators. In the case of 0-form symmetry, the symmetry defects act on local operators. This is because the configuration of local charged operators is topologically distinct inside and outside the symmetry defect. In fact, consider the 0-form symmetry defect $U(M)$ on the boundary of the region D, $(\partial D = M)$. If there are no local operators on the region D, we can remove the defects. But there exist local operators on D, we can not remove the defects without acting on these. but $p(>0)$-form symmetry defects is codimension-$(p+1)$ and does not bisect space. In any case, we can remove p-form symmetry defects without crossing local operators. In general, a p-form symmetry only acts on the operators of dimension $q > p$. In this paper, we only consider the

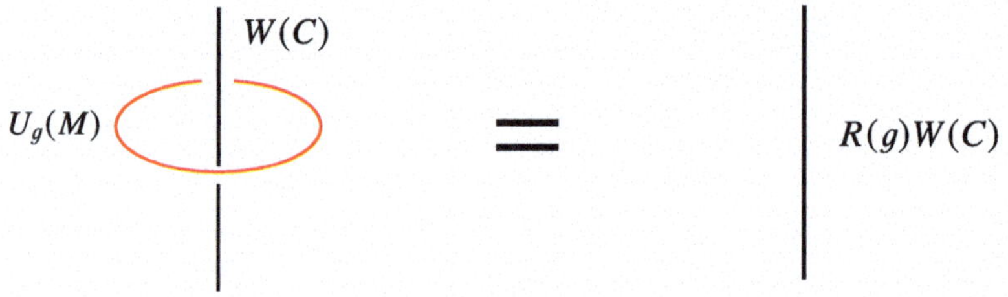

Fig. 2.11 Action of p-form symmetry defects. The red circle represents a p-form symmetry defect $U_g(M)$ on the $d-p-1$-dimensional surface M. The black line represents a charged operator $W(C)$ on the p dimensional surface C. The p-form symmetry defect $U_g(M)$ acts by linking to a charge operator $W(C)$

case of $q = p$. The p-form symmetry defect $U_g(M)$ act on the charged operator $W(C)$ on the p-dimensional surface C as follow,

$$U_g(M)W(C) = R(g)W(C), \tag{2.2.4}$$

where the M and C are linked in spacetime as in Fig. 2.11 and the R is representation of the g.

2.2.1 Example: 4-Dimensional $\mathbb{Z}_2$ Lattice Gauge Theory

In this subsection, we consider 4-dimensional $\mathbb{Z}_2$ lattice gauge theory and 1-form $\mathbb{Z}_2$ symmetry defect. We consider a 4-dimensional hyper cubic lattice system. Let us denote the plaquette consisting the sites i, j, k, l on its boundary by $\langle ijkl\rangle$. We assign the link variable $a_{ij} = 0, 1$ to the link $\langle ij\rangle$ and interaction

$$K(-1)^{a_{ij}+a_{jk}+a_{kl}+a_{li}} \tag{2.2.5}$$

to the plaquette $\langle ijkl\rangle$. The K is a positive constant parameter.

The partition function of the 4-dimensional $\mathbb{Z}_2$ lattice gauge theory is

$$Z = \sum_{\{a\}} \exp\left\{ \sum_{\substack{\text{plaquette} \\ \langle ijkl\rangle}} K(-1)^{a_{ij}+a_{jk}+a_{kl}+a_{li}} \right\}. \tag{2.2.6}$$

This theory is invariant under the following gauge transformation.

$$a_{ij} \to a'_{ij} = a_{ij} + \lambda_i + \lambda_j \ (\text{mod } 2), \quad (\lambda_i = 0, 1, \ i : \text{site}), \tag{2.2.7}$$

where the $\lambda_i = 0, 1$ is a parameter of a gauge transformation.

We consider a set of links $\tilde{D}$ containing up to two links for each plaquette. And we define the boundary $\tilde{M}$ of $\tilde{D}$ as the set of plaquettes where only one of the links that constitute it belong to the set $\tilde{D}$. As an analogy to the 0-form partial global transformation, we consider a partial 1-form global transformation as follow,

$$a_{ij} \to a_{ij}^{-1;D} \equiv \begin{cases} a_{ij} + 1 \pmod 2 & (x \in \tilde{D}) \\ a_{ij} & (x \notin \tilde{D}) \end{cases}. \tag{2.2.8}$$

Let us calculate the variation of the action. By performing this transformation, the interaction of the plaquette $\langle ijkl \rangle$ changes as follow,

$$K(-1)^{a_{ij}+a_{jk}+a_{kl}+a_{li}} \to \begin{cases} -K(-1)^{a_{ij}+a_{jk}+a_{kl}+a_{li}} & (\langle ijkl \rangle \in \tilde{M}) \\ K(-1)^{a_{ij}+a_{jk}+a_{kl}+a_{li}} & (\langle ijkl \rangle \notin \tilde{M}) \end{cases}. \tag{2.2.9}$$

In other words, only the plaquettes belonging to $\tilde{M}$ contributes to the variation of the action.

Using this variation, we can construct a symmetry defect. Let's check where this symmetry defect exists. We can construct the dual lattice putting dual sites at the center of the all hypercubes in original lattice system. Then in dual lattice, we define a dual cube $\widetilde{\langle ij \rangle}$ consisting of eight dual sites, with original link $\langle ij \rangle$ orthogonal to its center. Similarly, we define a dual plaquette $\widetilde{\langle ijkl \rangle}$ in dual lattice with original plaquette $\langle ijkl \rangle$ orthogonal to its center. The sets D and M are defined as follow

$$D \equiv \{\widetilde{\langle ij \rangle} | \langle ij \rangle \in \tilde{D}\}, \tag{2.2.10}$$

$$M \equiv \{\widetilde{\langle ijkl \rangle} | \langle ijkl \rangle \in \tilde{M}\}. \tag{2.2.11}$$

The projection of the sets $\tilde{D}$, $\tilde{M}$, D, and M in 2 dimensions is Fig. 2.12. Figure 2.13 shows the sets D and M projected in three dimensions. The set M constitute the 2-dimensional surface without boundary. We can consider a symmetry defect $U_{-1}(M)$ to exist on M.

Next we consider the action of 1-form $\mathbb{Z}_2$ symmetry defect on a charged operator. The charge operator is a Wilson loop

$$W(C) = \prod_{l \in C} (-1)^{a_l}, \tag{2.2.12}$$

where the C is the set of links that form a closed loop. The transformation $a^{-1;D}$ flip only the variable a_l of the link l crossing across D, this means that the link l belong to the $\tilde{D}$. For simplicity, let us assume that there is only one link in the set C that crosses M. The following WT-like identity holds:

$$\langle U_{-1}(M) W(C) \rangle = -\langle W(C) \rangle. \tag{2.2.13}$$

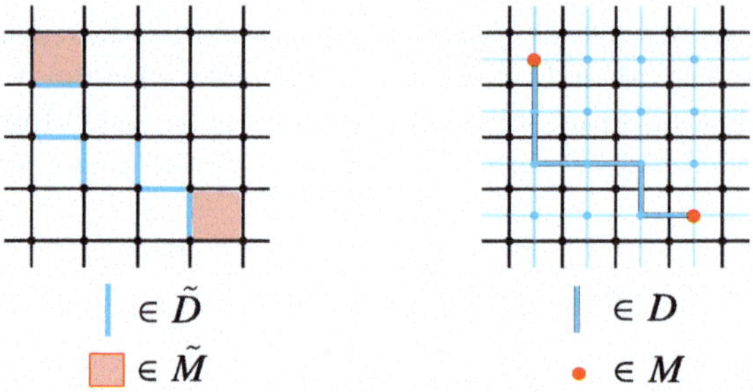

Fig. 2.12 Schematic illustration of 4-dimensional lattices. In the left figure, the blue lines represent links belonging to $\tilde{D}$ and the red squares represent plaquettes belonging to $\tilde{M}$. In the right figure, the blue lines bordered in black represent dual cubes to the blue links in the left figure. These dual cubes belong to D. The red dots represent plaquettes dual to the red plaquette in the left figure, and these belong to M. In this figure, M and $\tilde{M}$ appear to be unconnected, but in actual 4-dimensional lattice, it is connected and has no boundary

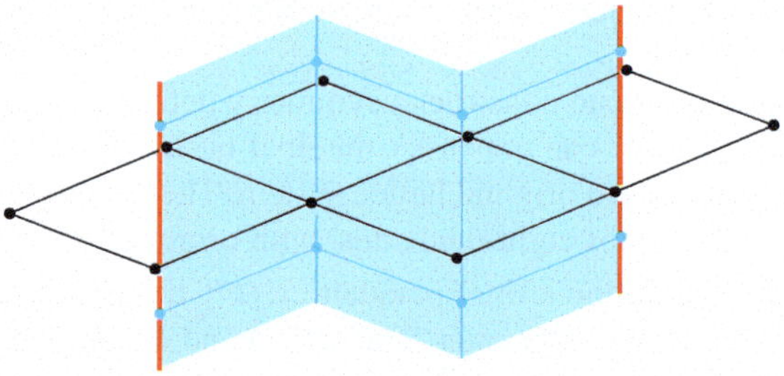

Fig. 2.13 Schematic illustration of the 4-dimensional lattices. In this figure, the vertical direction is projected to have 2-dimensional degrees of freedom. The black links and dots form the original lattice. Blue dots and blue links form the dual lattice. The blue and red vertical lines represent dual plaquettes, and the blue diagonal lines form a dual-pair link. The blue surface represents the 3-dimensional region D consisting of dual cubes. The red line represents the 2-dimensional surface M of the boundary of D

2.3 Non-invertible Symmetry

All the symmetry defects we have considered so far have a group structure. That is, the operation of the group is reproduced by fusion.

$$U_{g_1} \times U_{g_2} = U_{g_1 g_2} \quad (g_1, g_2 \in G). \tag{2.3.1}$$

This operation satisfies associativity.

$$(U_{g_1} \times U_{g_2}) \times U_{g_3} = U_{g_1} \times (U_{g_2} \times U_{g_3}) \quad (g_1, g_2, g_3 \in G). \tag{2.3.2}$$

There exists an identity defect **1**. And in particular, for any defect there exists its inverse defect. For any defect U_g, there exists an inverse defect $U_g^{-1} = U_{g^{-1}}$, and they generated the identity defect by fusion.

$$U_g \times U_{g^{-1}} = \mathbf{1} \quad (g, g^{-1} \in G). \tag{2.3.3}$$

Topological defects that have this invertible property are called invertible defects.

On the other hand, general topological defects do not necessarily have an inverse defect. These defects is called non-invertible defects. Specific examples of non-invertible symmetry and defect are discussed in the following chapters.

Chapter 3
Ising Model and Kramers-Wannier Duality

The purpose of this chapter is to introduce one of the approaches to constructing non-invertible topological defects, the Aasen-Mong-Fendley' approach. The 2-dimensional Ising model is dual to the gauged theory. This duality is called KW duality. Aasen-Mong-Fendley showed that non-invertible topological defects can be constructed as an interface between those dual theories.

In this section, we first verify that KW duality holds. We define defects on the Ising model introducing the dual lattice and impose conditions for these defects to be topological. In particular, we confirm the existence of solutions for non-invertible defects that induce KW duality transformations (KW duality defects). We also investigate the fusion of KW defects and spin-flip $\mathbb{Z}_2$ symmetry defects.

3.1 Kramers-Wannier Duality

In this section, we explain and verify the Kramers-Wannier duality. To provide a detailed explanation of this duality, we will specifically define the model. We consider a 2-dimensional square lattice system of N_s sites with periodic boundary condition. We assign the labels i, j to the sites and the variable $a_i = 0, 1$ to the site i. We define the partition function of 2-dimensional Ising model Z_{Ising} as

$$Z_{\text{Ising}}(K) = \sum_{\{a\}} \exp\left(K \sum_{\substack{\text{link} \\ \text{l}=\langle ij\rangle}} (-1)^{(a_i+a_j)}\right) \tag{3.1.1}$$

$$= \sum_{\{a\}} \prod_{\substack{\text{link} \\ \text{l}=\langle ij\rangle}} \exp\left(K(-1)^{(a_i+a_j)}\right), \tag{3.1.2}$$

M. Koide, *Non-invertible Symmetry in 4-Dimensional Z2 Lattice Gauge Theory*, Springer Theses, https://doi.org/10.1007/978-981-96-2272-6_3

where the sum $\sum_{\{a\}}$ means the summation of the all configuration of the variables a, the K is positive constant parameter.

This theory has a 0-form global $\mathbb{Z}_2$ symmetry

$$a_i \to a_i' = a_i + 1 \pmod 2, \tag{3.1.3}$$

$$\exp\left(K(-1)^{(a_i'+a_j')}\right) = \exp\left(K(-1)^{(a_i+a_j)}\right). \tag{3.1.4}$$

We consider gauging this symmetry. The original Ising model $Z_{\text{Ising}}(K)$ is not invariant by the 0-form gauge $\mathbb{Z}_2$ transformation,

$$a_i \to a_i' = a_i + \lambda_i \pmod 2, \tag{3.1.5}$$

where the $\lambda_i = 0, 1$ is the parameter of the gauge transformation. To absorb this change, we introduce the gauge field b_l on the link l. We define the gauge transformation of a gauge field b_l, associated with the gauge transformation of a site, by

$$b_l \to b_l' = b_l + \lambda_{i_1(l)} + \lambda_{i_2(l)}, \tag{3.1.6}$$

where the $i_1(l), i_2(l)$ are the sites in the link l. Then, to make the partition function invariant under the gauge transformation, the Boltzmann weights are rewritten as in

$$\exp\left(K(-1)^{(a_{i_1(l)}+a_{i_2(l)}+b_l)}\right). \tag{3.1.7}$$

We also introduce the action of the gauge field as

$$\frac{1}{2}\sum_{c_p=0,1} e^{\pi i c_p (b_{l_1(p)}+b_{l_2(p)}+b_{l_3(p)}+b_{l_4(p)})}, \tag{3.1.8}$$

where p is a plaquette consisting of four links $l_1(p), l_2(p), l_3(p), l_4(p)$. This action takes the weak coupling limit for the usual Wilson's plaquette action. Two spin configurations connected by a gauge transformation imply the same configuration. Therefore, to count these correctly, assign a weight of $1/2$ for each site. Then, we define the partition function of the $\mathbb{Z}_2$ symmetry gauged 2-dimensional ising model $Z_{\text{Ising}/\mathbb{Z}_2}(K)$ as

$$Z_{\text{Ising}/\mathbb{Z}_2}(K) = \sum_{\{a\}}\sum_{\{b\}}\sum_{\{c\}} \frac{1}{2^{2N_s}} \exp\left(K\sum_{l:\text{link}}(-1)^{(a_{i_1(l)}+a_{i_2(l)}+b_l)} + i\pi \sum_{p:\text{plaquette}} c_p(b_{l_1(p)}+b_{l_2(p)}+b_{l_3(p)}+b_{l_4(p)})\right). \tag{3.1.9}$$

Also, by choosing a gauge constraint of $a_i = 0$, we obtain the following partition function,

$$Z_{\text{Ising}/\mathbb{Z}_2}(K) = \sum_{\{b\}} \sum_{\{c\}} \frac{1}{2^{N_s}}$$
$$\exp\left(K \sum_{l:\text{link}} (-1)^{b_l} + i\pi \sum_{p:\text{plaquette}} c_p(b_{l_1(p)} + b_{l_2(p)} + b_{l_3(p)} + b_{l_4(p)}) \right). \tag{3.1.10}$$

The Kramers-Wannier duality is a relationship between 2-dimensional Ising model $Z_{\text{Ising}}(K)$ and $\mathbb{Z}_2$ symmetry gauged 2-dimensional Ising model. Let $Z_{\text{Ising}/\mathbb{Z}_2}(K)$ be the partition function of the $\mathbb{Z}_2$ symmetry gauged 2-dimensional Ising model. Then the Kramers-Wannier duality is the relation as follow,

$$\frac{1}{(\sinh 2K)^{N_s/2}} Z_{\text{Ising}}(K) = \frac{1}{(\sinh 2\tilde{K})^{N_s/2}} Z_{\text{Ising}/\mathbb{Z}_2}(\tilde{K}), \tag{3.1.11}$$
$$\sinh 2K \sinh 2\tilde{K} = 1. \tag{3.1.12}$$

We deform Eq. (2.1.33) to prove the Kramers-Wannier duality Eq. (3.1.11). First, we define the dual constant parameters $\tilde{K}$ as

$$\tanh \tilde{K} = e^{-2K}. \tag{3.1.13}$$

This equation satisfies Eq. (3.1.12).

$$\begin{aligned} \sinh 2K &= \frac{1}{2}\left(e^{2K} - e^{-2K}\right) \\ &= \frac{1}{2}\left(\frac{1}{\tanh \tilde{K}} - \tanh \tilde{K}\right) \\ &= \frac{1}{2}\left(\frac{\cosh^2 \tilde{K} - \sinh^2 \tilde{K}}{\sinh \tilde{K} \cosh \tilde{K}}\right) \\ &= \frac{1}{\sinh 2\tilde{K}}. \end{aligned} \tag{3.1.14}$$

From Eq. (3.1.13), we obtain the relation,

$$\begin{aligned} e^{K(-1)^a} &= \frac{\cosh \tilde{K} (\tanh \tilde{K})^a}{\sqrt{\cosh \tilde{K} \sinh \tilde{K}}} \\ &= \frac{2\cosh \tilde{K} (\tanh \tilde{K})^a}{\sqrt{2 \sinh 2\tilde{K}}} \\ &= \frac{1}{\sqrt{2 \sinh 2\tilde{K}}} \sum_{b=0,1} \exp \tilde{K}(-1)^b + i\pi ab \quad (a = 0, 1). \end{aligned} \tag{3.1.15}$$

We substitute this equation for Eq. (2.1.33).

$$
\begin{aligned}
Z_{\text{Ising}}(K) &= \frac{1}{(2\sinh 2\tilde{K})^{N_l/2}} \sum_{\{a\}} \prod_{l:\text{link}} \left(\sum_{b_l=0,1} \exp(K(-1)^{b_l} + i\pi(a_{i_1(l)} + a_{i_2(l)})b_l) \right) \\
&= \frac{1}{2^{N_s}(\sinh 2\tilde{K})^{N_s}} \sum_{\{a\}} \sum_{\{b\}} \left(\exp\left(\sum_{l:\text{link}} (K(-1)^{b_l} + i\pi(a_{i_1(l)} + a_{i_2(l)})b_l) \right) \right), \quad (3.1.16)
\end{aligned}
$$

where $N_l = 2N_s$ is the number of the link in this square lattice. We convert the sum for the link variable b_l taken in the second term of the exponential power to a sum for each site variable a_i. Since each site has four links connected to it, we can transform thia equation as in

$$
\begin{aligned}
Z_{\text{Ising}}(K) =& \frac{1}{2^{N_s}(\sinh 2\tilde{K})^{N_s}} \\
& \sum_{\{a\}} \sum_{\{b\}} \exp\left(\sum_{l:\text{link}} K(-1)^{b_l} + i\pi \sum_{i:\text{site}} a_i(b_{l_1(i)} + b_{l_2(i)} + b_{l_3(i)} + b_{l_4(i)}) \right).
\end{aligned}
\quad (3.1.17)
$$

Now prepare a lattice of links that are orthogonal to the midpoint of each link in the original lattice. At this time, the site on the original lattice becomes the plaquette of the new lattice. Therefore, we redefine site variable a_i in the original lattice as a plaquette variable $\tilde{c}_i$ in the new lattice. It also defines the link variables $\tilde{b}_l$ in the new lattice by the link variables b_l the original lattice, which are in an orthogonal relationship to each other. Thus we prove the Kramers-Wannier duality Eq. (3.1.11).

Let's focus on when $K = \tilde{K} = K_c$ is satisfied. Substituting this equation for (3.1.12), we find that $K_c = \frac{1}{2}\log(1+\sqrt{2})$. This parameter K_c indicates the critical point. In this case, Kramers-Wannier duality Eq. (3.1.11) becomes as

$$
Z_{\text{Ising}}(K_c) = Z_{\text{Ising}/\mathbb{Z}_2}(K_c). \quad (3.1.18)
$$

This kind of relationship is called self-duality.

3.2 Dual Lattice

To construct duality defects, we introduce two types of lattices, as illustrated in Fig. 3.1. To provide a more detailed description, we introduce a two-dimensional coordinate system, denoted as (x_1, x_2) in $\mathbb{R}^2$. Points with Λ have coordinates consisting of even integers, defined as $\Lambda := \{(x_1, x_2) | x_1, x_2 \in 2\mathbb{Z}\}$. Conversely, for $\hat{\Lambda}$, the coordinates are formed by odd integers, expressed as $\hat{\Lambda} := \{(x_1, x_2) | x_1, x_2 \in 2\mathbb{Z} + 1\}$.

We assign the site variables $U_i = (-1)^{a_i} = \pm 1$ to the sites in Λ, while we do not assign site variables to the sites in $\hat{\Lambda}$ For this reason, we refer to the lattice Λ, to which the site variables are assigned, as the active lattice. We label the sites on the

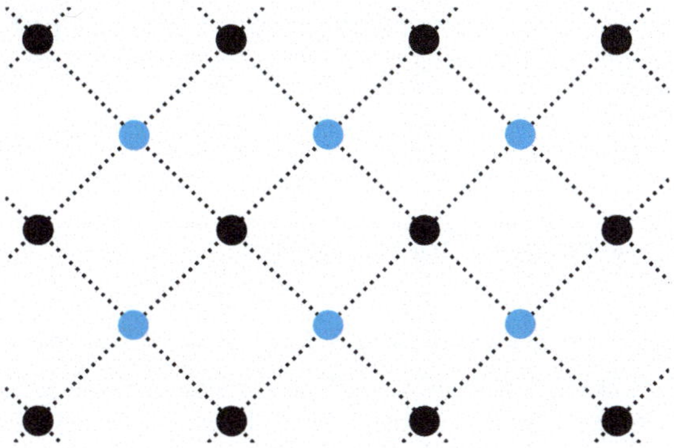

Fig. 3.1 Illustration of a 2-dimensional lattice in the AMF approach. Black dots represent active sites to which spin variables are assigned. Blue dots represent inactive sites with no spin variable assigned. Each square composed by four dotted lines contains two active sites and two inactive sites. This square is called the Boltzmann plaquette. Reprinted from [1], Copyright 2021, The Author(s) licensed under CC BY 4.0

lattice Λ as active sites. On the other hand, the lattice $\hat{\Lambda}$ is referred to as the inactive lattice. We label the sites on the lattice Λ as inactive sites. We assign the weights $\tilde{s}$ to the inactive sites. The specific value of this weight is determined later. We also use the term Boltzmann plaquette for the smallest square formed by this lattice.

Let $a_i = 0, 1 (i = 1, 2)$ be the site variables assigned to the two closest sites to each other in active lattice. We define the Boltzmann weight of the Ising model as

$$W(a_1, a_2) = \exp(K(-1)^{(a_1+a_2)}). \tag{3.2.1}$$

There is one-to-one correspondence between two closest sites to each other in active lattice and the Boltzmann plaquette containing these sites. So, we can consider the basic unit of this lattice system as a plaquette.

We define the partition function of the Ising model as

$$Z_{\text{Ising}}(K) = \sum_{\{a\}} \left(\prod_{\substack{\text{inactive} \\ \text{sites}}} \tilde{s} \right) \exp\left(K \sum_{p:\text{plaquette}} (-1)^{(a_{i_1(p)}+a_{i_2(p)})} \right). \tag{3.2.2}$$

Besides the constant normalization $\prod s$, it is identical to the partition function (2.1.33).

3.3 KW Duality Defect in 2-Dimensional Ising Model

In this section, we construct the KW duality defects, $\mathbb{Z}_2$ symmetry defects and junctions among them following Aasen-Mong-Fendley's approach ([2]). We impose on a defect or junction equations that a topological defect must satisfy. We determine the weights by solving their equations. We also show that the duality defects are non-invertible symmetry.

3.3.1 Kramers-Wannier Duality Defect

In this subsection, we discuss the duality defects in the 2-dimensional Ising model. Duality defects are line operators and are placed at the boundary of the two regions associated with KW duality. Let's review the discussion in Sect. 3.1. In the KW duality, the sites of the original Ising model lattice are reinterpreted as plaquettes of the gauged Ising model lattice and vice versa. Using the terms active lattice and inactive lattice, this correspondence is expressed by the relationship between active sites and inactive sites. Therefore, we define duality defects on the lattice to swap active and inactive sites, as in Fig. 3.2. The building block of the duality defects is a parallelogram with 2 active sites in Fig. 3.2. Since the building block contains two active sites i, j, we assign weight $D(a_i, a_j)$ to the building block.

Since the duality transformation only changes the description of the theory and does not change the observables, we expect the existence of topological operators on the transformation. In the following, we describe the commutation relations necessary for the duality defects to be topological, and we construct a topological duality defect by solving its equations.

In our setup, the Boltzmann plaquettes fill the lattice. Therefore, let us focus on a single Boltzmann plaquette and consider commutation relations. On a single Boltzmann plaquette, it is possible to consider two kinds of commutation relations, topological deformations of defects, as represented in Figs. 3.3 and 3.4.

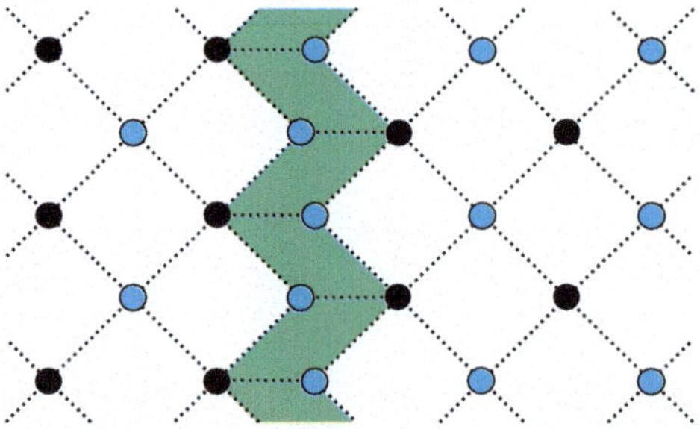

Fig. 3.2 Illustration of duality defects on two-dimensional lattice in the AMF approach. Black and blue dots represent active and inactive sites, respectively. The green parallelograms containing two active and two inactive sites each represent building blocks of duality defects. Duality defects are placed at the boundaries of mutually dual theories and swap active and inactive sites.

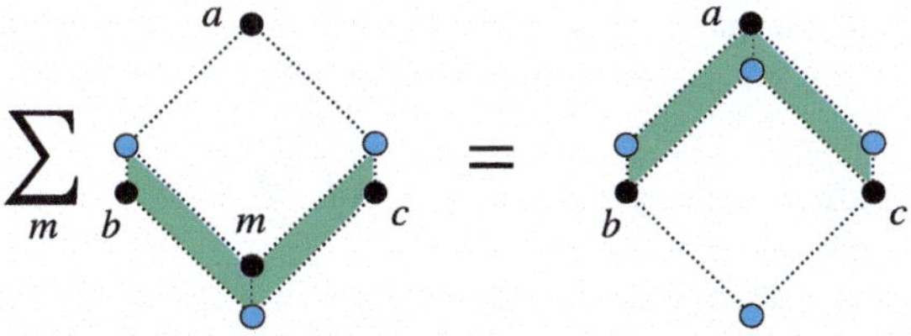

Fig. 3.3 One of the duality defect commutation relations. Black and blue dots represent active and inactive sites, respectively. The green parallelograms represent the building blocks of the duality defects. This equality imposes that a duality defect can be topologically deformed from two building blocks to two building blocks

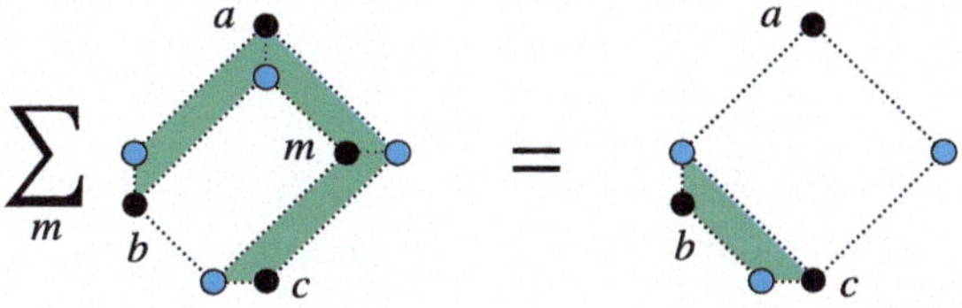

Fig. 3.4 One of the duality defect commutation relations. Black and blue dots represent active and inactive sites, respectively. The green quadrangles represent the building blocks of the duality defects. This equality imposes that a duality defect consisting of three building blocks can be topologically deformed into a duality defect consisting of one building block, and vice versa

These are represented as equations

$$\sum_{m=0,1} W(a,m)D(b,m)D(c,m)\tilde{s}^3 = W(b,c)D(b,a)D(c,a)\tilde{s}^4, \tag{3.3.1}$$

$$\sum_{m=0,1} W(a,m)D(a,b)D(a,m)D(c,m)\tilde{s}^4 = W(a,c)D(b,c)\tilde{s}^3. \tag{3.3.2}$$

If these equations hold on a single Boltzmann plaquette, the whole defect becomes topological by repeating the deformation on each Boltzmann plaquette.

We determine the values $D(a_i, a_j)$, $\tilde{s}$, K by solving these defect commutation relations. In a physically sensible solution, these values satisfy

$$D(a,\tilde{a}) \neq 0, \quad \tilde{s} > 0, \quad K \in \mathbb{R}, \quad K \neq 0. \tag{3.3.3}$$

There is a unique physically sensible solution up to the sign of $D(a,\tilde{a})$. The solution is

$$D(a,\tilde{a}) = \frac{1}{\sqrt{2}}(-1)^{a\tilde{a}}, \tag{3.3.4}$$

$$\tilde{s} = \sqrt{2}, \tag{3.3.5}$$

$$K = K_c = -\frac{1}{2}\log(-1+\sqrt{2}), \qquad W(a_i,a_j) = \exp(K(-1)^{(a_i+a_j)}). \tag{3.3.6}$$

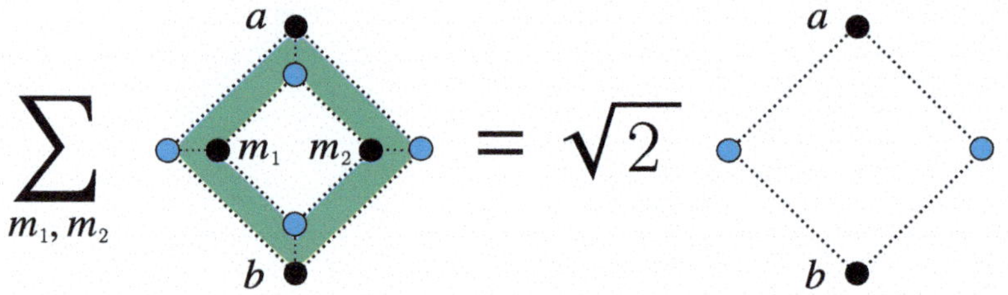

Fig. 3.5 Calculation of S^1 expectation value of duality defect. Black and blue dots represent active and inactive sites, respectively. The green quadrangles represent the building blocks of the duality defects. In the left figure, the duality defects are placed on each link that constitutes the Boltzmann plaquette, and the duality defects have an S^1 topology. This coordination is equal to $\sqrt{2}$ times that of the Boltzmann packet without duality defects

Note that the value of K is determined by the critical value K_c. The sign ambiguity of $D(a_i, a_j)$ does not affect the observables, at least locally. We discuss a bit more on this sign ambiguity in Sect. 6

We prove that this KW duality is a non-invertible symmetry. We consider a defect configuration that doubles all four links of a single plaquette. At this time, the defect forms a closed manifold S^1, as shown in Fig. 3.5. Substituting and calculating each weight, we find that this coordination is proportional to the Boltzmann plaquette where no defect is placed, as shown in

$$\sum_{m_1, m_2=0,1} W(m_1, m_2) D(a, m_1) D(a, m_2) D(b, m_1) D(b, m_2) \tilde{s}^4 = \sqrt{2} W(a, b) \tilde{s}^2. \tag{3.3.7}$$

In the case of a symmetry defect placed on a closed manifold with no operator inserted inside, the weight is the same as the empty configuration. Therefore, we conclude that the KW duality defects are non-invertible topological defects.

3.3.2 $\mathbb{Z}_2$ Symmetry Defects in the 2d Ising Model

The Ising model has $\mathbb{Z}_2$ spin flip symmetry. Next we construct the $\mathbb{Z}_2$ symmetry defects on the our setup. The defects are line operators. Since the $\mathbb{Z}_2$ symmetry is a spin-flipping symmetry, we consider a line defects that doubles active sites to active sites and inactive sites to inactive sites, as in Fig. 3.6. The building block of the symmetry defects is a parallelogram in Fig. 3.6. Since the building block contains two active sites i, j, we assign weight $Z(a_i, a_j)$ to the building block.

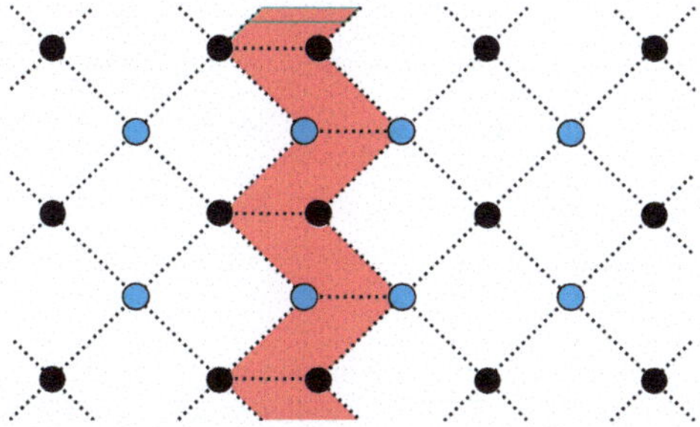

Fig. 3.6 Illustration of a $\mathbb{Z}_2$ symmetry defect on two-dimensional lattice in the AMF approach. Black and blue dots represent active and inactive sites, respectively. The red parallelograms containing two active and two inactive sites each represent building blocks of $\mathbb{Z}_2$ symmetry defects. Modified from [1], Copyright 2021, The Author(s) licensed under CC BY 4.0

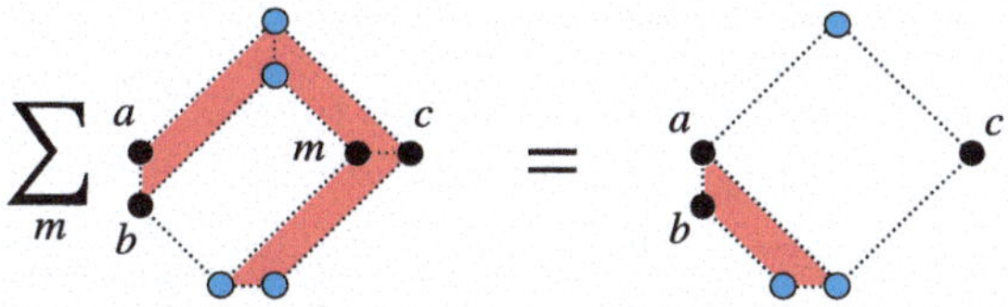

Fig. 3.7 One of the $\mathbb{Z}_2$ symmetry defect commutation relations. Black and blue dots represent active and inactive sites, respectively. The red quadrangles represent the building blocks of the $\mathbb{Z}_2$ symmetry defects. This equality imposes that a $\mathbb{Z}_2$ symmetry defect consisting of three building blocks can be topologically deformed into a $\mathbb{Z}_2$ symmetry defect consisting of one building block, and vice versa

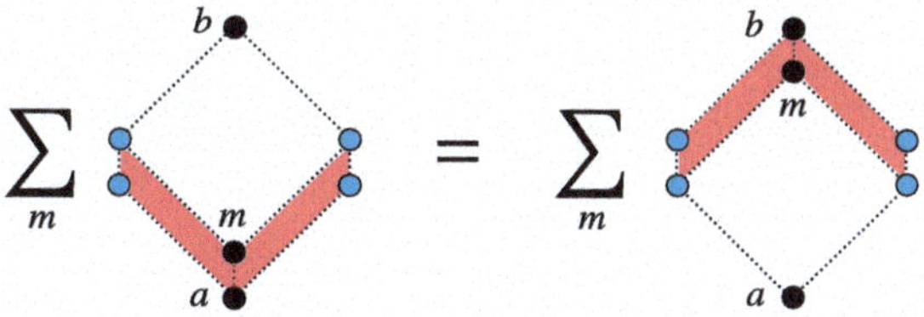

Fig. 3.8 One of the $\mathbb{Z}_2$ symmetry defect commutation relations. Black and blue dots represent active and inactive sites, respectively. The red parallelograms represent the building blocks of the $\mathbb{Z}_2$ symmetry defects. This equality imposes that a $\mathbb{Z}_2$ symmetry defect can be topologically deformed from two building blocks to another two building blocks

As in the case of the KW duality defects, we focus on a single Boltzmann plaquette and consider defect commutation relations. In the single Boltzmann plaquette we can consider three kinds of defect commutation relations for symmetry defects, as Figs. 3.7, 3.8 and 3.9. There are represented as equations

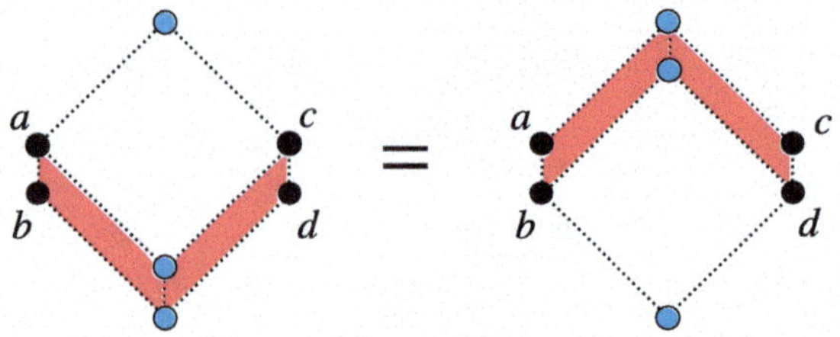

Fig. 3.9 One of the $\mathbb{Z}_2$ symmetry defect commutation relations. Black and blue dots represent active and inactive sites, respectively. The red parallelograms represent the building blocks of the $\mathbb{Z}_2$ symmetry defects. This equality imposes that a $\mathbb{Z}_2$ symmetry defect can be topologically deformed from two building blocks to another two building blocks

$$\sum_{m=0,1} W(b,m)Z(a,b)Z(c,m)Z(c,m)\tilde{s}^4 = W(a,b)Z(a,b)\tilde{s}^3, \tag{3.3.8}$$

$$\sum_{m=0,1} W(b,m)Z(a,m)^2\tilde{s}^4 = \sum_{m=0,1} W(a,m)Z(b,m)^2\tilde{s}^4, \tag{3.3.9}$$

$$W(a,c)Z(a,b)Z(c,d)\tilde{s}^3 = W(b,d)Z(a,b)Z(c,d)\tilde{s}^3. \tag{3.3.10}$$

Equation (3.3.10) is a identity. We determine the weights $Z(a,b)$ by solving these equations. Since we want to consider models for which duality defects can be defined, we impose condition

$$\tilde{s} = \sqrt{2}, \quad K = K_c = -\frac{1}{2}\log(-1+\sqrt{2}). \tag{3.3.11}$$

In a physically sensible solution, the weights satisfies condition,

$$Z(a_i, a_j) \neq 0. \tag{3.3.12}$$

Then we find the two solutions up to sign

$$Z(a_i, a_j) = 2^{-\frac{1}{4}}(1 - \delta_{a_i,a_j}), \tag{3.3.13}$$

$$Z(a_i, a_j) = 2^{-\frac{1}{4}}\delta_{a_i,a_j}. \tag{3.3.14}$$

The solution (3.3.13) means that the spins of the two active sites connected by the defects are different. Therefore, this solution is the $\mathbb{Z}_2$ symmetry defects. On the other hand, the solution (3.3.14) means that the spins of the two active sites connected by the defects are same. this solution corresponds to the trivial defects. We again define $Z(a,b)$ as the weights of the basic unit of the $\mathbb{Z}_2$ symmetry defects and $T(a,b)$ as the weights of the basic unit of the trivial defects.

These topological defects are invertible. The weight of the Boltzmann plaquette with these defects placed on each link is equal to the weight of the empty Boltzmann plaquette.

$$\sum_{m_1,m_2=0,1} W(m_1,m_2)Z(a,m_1)^2 Z(b,m_2)^2 \tilde{s}^4 = W(a,b)\tilde{s}^2. \tag{3.3.15}$$

$$\sum_{m_1,m_2=0,1} W(m_1,m_2)T(a,m_1)^2 T(b,m_2)^2 \tilde{s}^4 = W(a,b)\tilde{s}^2. \tag{3.3.16}$$

3.3.3 Topological Defect Junction

We can consider the fusion of topological defects by smoothly deforming and superimposing defects. By fusion, junctions can arise where multiple topological line operators intersect at a single point. A junction that can move freely as long as it does not intersect with other defects is said to be topological. In this section, we construct the topological trivalent junctions, intersected with three defects, on the our setup.

We want to construct topological trivalent junctions. To do so, we need to impose commutation relations on the junction that must be satisfied in order for the junction to be topological. Now, we consider the three topological line defects R, G, B and their junction as shown in the left figure in Fig. 3.10. As shown in the right figure in Fig. 3.10, when these defects are defined by doubling the site on the lattice, as in the KW defect, $\mathbb{Z}_2$ symmetry defects, the junction can then be represented by a triangle on this lattice as shown in Fig. 3.11. In Fig. 3.11, the α, β, γ are labels of the sites.

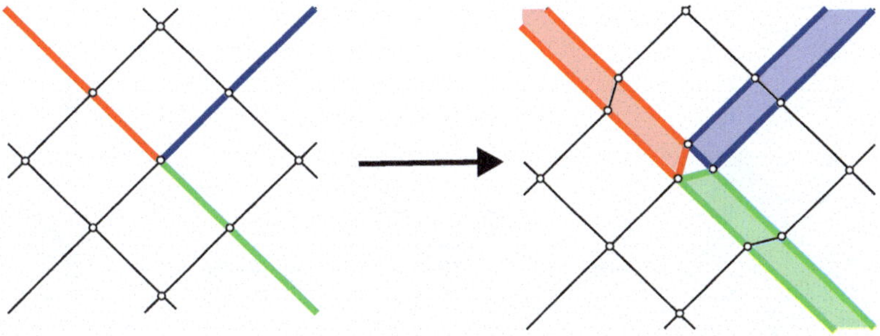

Fig. 3.10 Illustration of three topological lines intersecting at a single junction. Left: red, blue, and green lines represent topological lines. Each line intersects at a single junction. Right: Illustration of topological lines on the lattice in the AMF approach. The colored quadrangles represent building blocks with topological defects. The triangles consisting of red, green, and blue edges represent the junctions of those topological lines

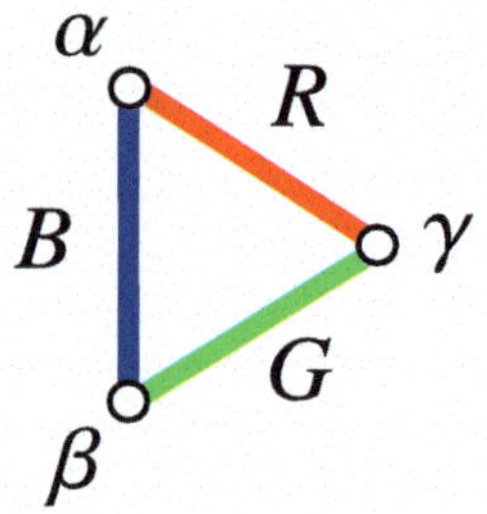

Fig. 3.11 Junction of three topological lines R, G, B. Red, green and blue edges are connected to topological lines R, G, and B, respectively. White vertices α, β, and γ represent the sites of the lattice in the AMF approach

If the site α is an active site, then the spin variable $U(a_\alpha) = (-1)^{a_\alpha}$ $(a_\alpha = 0, 1)$ is assigned to that site as a weight, and if it is an inactive site, then $\tilde{s}$ is assigned as a weight.

Next, we check what commutation relations are imposed on the junction in order to move the junction freely. We consider the movement of a junction as in Fig. 3.12. The top two figures in Fig. 3.12 represent the movement of the junction on the continuous theory. The bottom three figures in Fig. 3.12 represent how the move is represented on

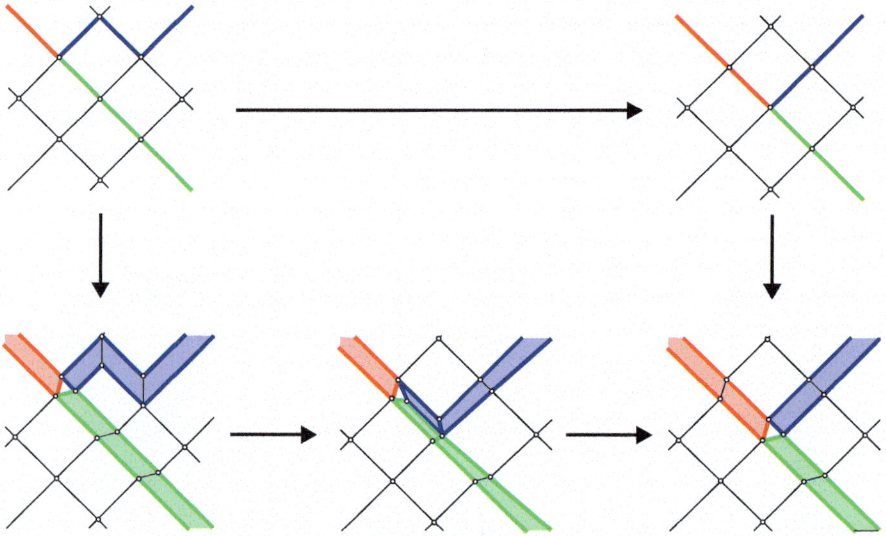

Fig. 3.12 Illustration of moving a junction topologically. The top row shows a conceptual illustration of topological line deformation and moving the junction, while the bottom row shows the actual deformation in the lattice in the AMF approach. In the first arrow of the bottom row, the defect is deformed using the defect commutation relation from two building blocks to another two building blocks for the blue defect. In the second arrow, the junction is moved to the lower right site by imposing Eq. (3.3.17)

the lattice. The deformation on the lattice is a two-step process. The first deformation uses the commutation relations for the topological defects R. In the second step, assuming the junction is topological, we move the junction to the right down. By this deformation, we get the following equation as the commutation relations imposed for a junction to be topological.

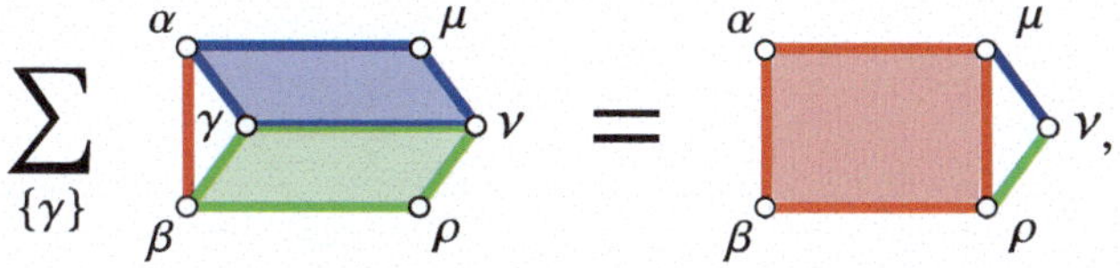

(3.3.17)

where the sum $\sum_{\{\gamma\}}$ means adding up for all possible weights of the site γ. By the deformation, the site γ disappears from the lattice, so the sum $\sum_{\{\gamma\}}$ is necessary. Thus, we find a general junction commutation relation.

Now we consider the junctions when the KW defects are placed as topological defect G, B and the $\mathbb{Z}_2$ symmetry defects as R. From Eq. (3.3.17), we obtain the two commutation relations imposed on these junctions,

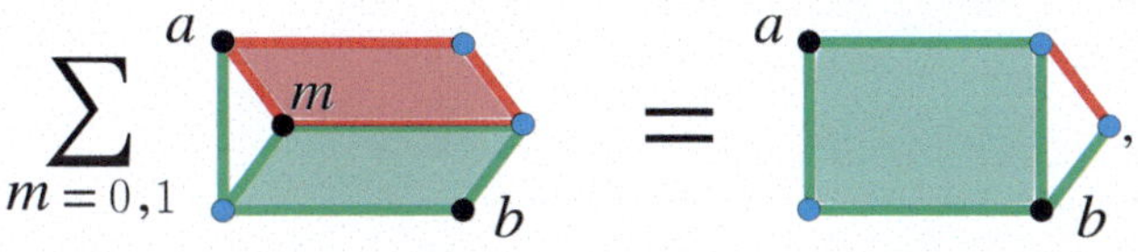

(3.3.18)

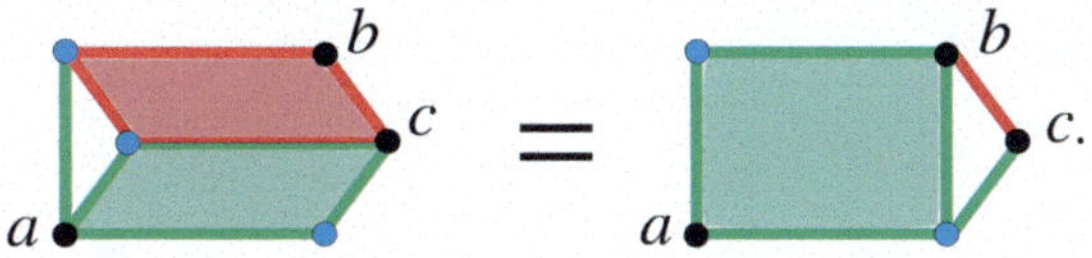

(3.3.19)

Then, there are two kinds of junction and we define these weights as $J_{KKZ}(a)$ and $J_{KKZ}(a_i, a_j)$, respectively, as follows.

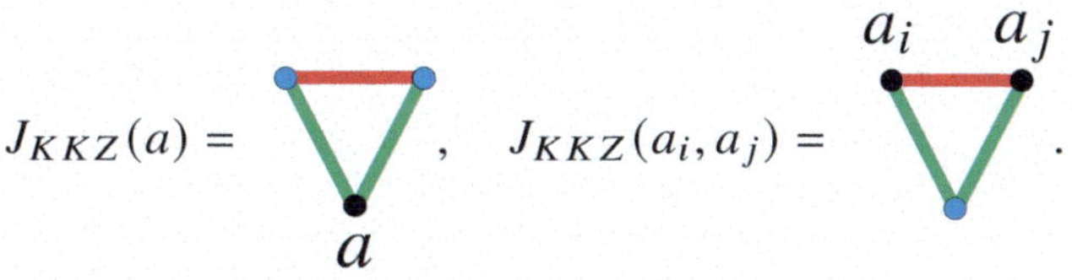

(3.3.20)

Using these weights $J_{KKZ}(a)$, $J_{KKZ}(a_i, a_j)$, the Eqs. (3.3.18), (3.3.19) are expressed in the formulas as equations

$$\sum_m J_{KKZ}(a,m)D(m,b)Z(a,m)\tilde{s}^3 = D(a,b)J_{KKZ}(b)\tilde{s}^3, \tag{3.3.21}$$

$$J_{KKZ}(a)D(a,b)Z(b,c)\tilde{s}^3 = D(a,b)J_{KKZ}(b,c)\tilde{s}^2. \tag{3.3.22}$$

By solving these equations we determined the solution up to overall factor J. The solution is as follows.

$J_{KKZ}(a) =$ 

$= (-1)^a J,\quad J_{KKZ}(a_i, a_j) =$ 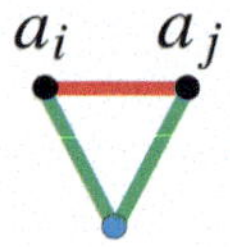

$= 2^{\frac{1}{4}}(1 - \delta_{a_i,a_j})J.$ (3.3.23)

To obtain the overall factor, we use the fact that the S^1 expectation value of the duality defects is $\sqrt{2}$. As shown in left figure in Fig. 3.13, consider a loop of the duality defects that contains no other operators inside and two $\mathbb{Z}_2$ symmetry defects that have the endpoints in the loop. Shrinking the loop of the duality defects to a point yields the S^1 expectation value $\sqrt{2}$ times the symmetry defects, as shown in Fig. 3.13. This is represented on the lattice as shown in Eq. (3.3.24).

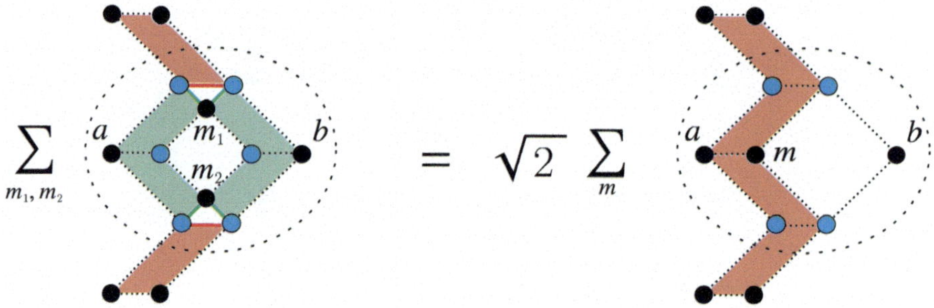

(3.3.24)

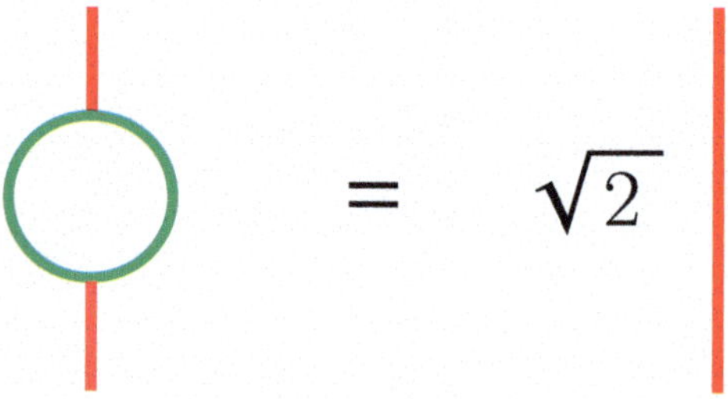

Fig. 3.13 Illustration of replacing the loop of the duality defect, where two $\mathbb{Z}_2$ symmetry defects are connected, by $\sqrt{2}$ times one $\mathbb{Z}_2$ symmetry defect. The red lines represent the $\mathbb{Z}_2$ symmetry defects. The green circle represents the duality defect. It is assumed that no operator is inserted in the loop

Since the outside of the dotted circle is not changed by the deformations of the defects, we only need to consider the single Boltzmann plaquette, the inside of the black dotted line. The Eq. (3.3.24) are expressed in the formulas as equations

$$\sum_{m_1,m_2} W(m_1, m_2)D(a, m_1)D(a, m_2)D(b, m_1)D(b, m_2)J_{KKZ}(m_1)J_{KKZ}(m_1)\tilde{s}^6$$
$$= \sqrt{2}\sum_m W(m, b)Z(m, a)^2\tilde{s}^4. \tag{3.3.25}$$

We determine $J = 2^{-\frac{1}{4}}$ by solving Eq. (3.3.25). Now we obtain the weights of junctions. These solutions are as follows.

$J_{KKZ}(a) =$

$= 2^{-\frac{1}{4}}(-1)^a, \quad J_{KKZ}(a_i, a_j) =$

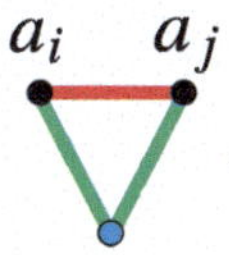

$= (1 - \delta_{a_i,a_j}).$ (3.3.26)

In the same way, we can also determine the weights of the junctions when the KW defects are placed as topological defects G, B and the trivial defects as R. Defining their weights as $J_{KKT}(a)$, $J_{KKT}(a_i, a_j)$, the values are

$J_{KKT}(a) =$

$= 2^{-\frac{1}{4}}, \quad J_{KKT}(a_i, a_j) =$

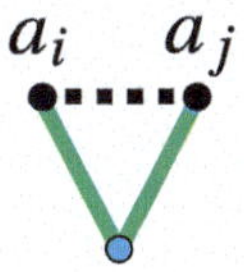

$= \delta_{a_i,a_j}.$ (3.3.27)

3.3.4 Crossing Relation

In the previous subsection, we have constructed the topological junctions. In this subsection, we use that junctions to represent a crossing relation, which are relations between different kinds of defects. We derive from them expectation values of the defects and fusion rules for defects on the 2-dimensional Ising model.

The first example of crossing relation is that $\mathbb{Z}_2$ symmetry defects connecting endpoints to duality defects can be removed if those endpoints are connected on the duality defects. It is shown schematically in Fig. 3.14. On the lattice, it is represented

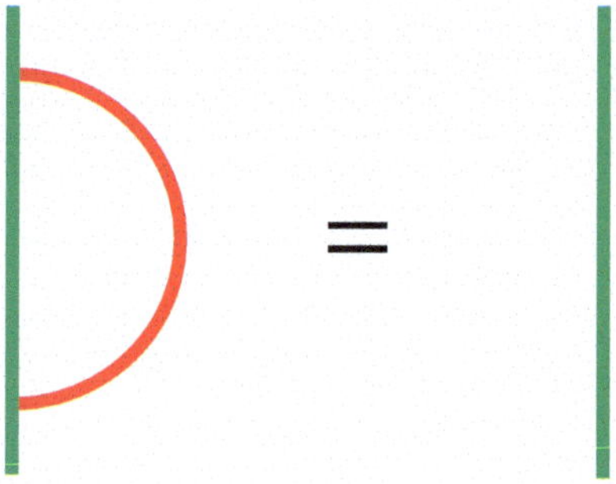

Fig. 3.14 $\mathbb{Z}_2$ symmetry defects connected to duality defects absorb into duality defects and disappear. The green and red lines represent the duality and $\mathbb{Z}_2$ symmetry defects, respectively. It is assumed that no operator is inserted in the region surrounded by each

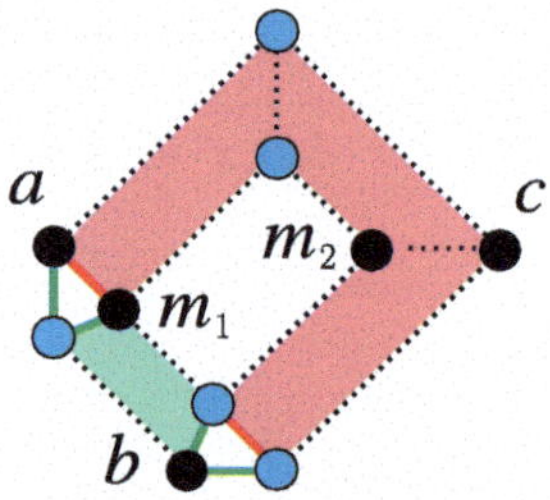

Fig. 3.15 Configuration of defects topologically equal to the left in Fig. 3.14 in the AMF approach lattice. Black and blue dots represent active and inactive sites, respectively. The green quadrangle represents the building block of the duality symmetry defects. The red quadrangles represent the building blocks of the $\mathbb{Z}_2$ symmetry defects

by the concrete example shown in Fig. 3.15. The crossing relation in this concrete example is as in Eq. (3.3.28). We can verify this equation.

$$\sum_{m_1,m_2} W(m_1,m_2)D(m_1,b)Z(a,m_1)Z(c,m_2)^2 J_{KKZ}(a,m_1)J_{KKZ}(b)\tilde{s}^5 = W(a,c)D(a,c)\tilde{s}^3. \tag{3.3.28}$$

Next, we consider a partial fusion of duality defects and $\mathbb{Z}_2$ symmetry defects. When $\mathbb{Z}_2$ symmetry defects are fused parallel to the duality defects, the fused $\mathbb{Z}_2$ symmetry defects are absorbed into the duality defects. It is shown schematically in Fig. 3.16. A concrete example on the lattice is Fig. 3.17, which is represented by Eq. (3.3.29). We can confirm that this equation holds.

$$W(a,c)D(a,b)Z(c,d)\tilde{s}^4 = \sum_m W(m,d)D(b,m)Z(a,m)Z(c,d)J_{KKZ}(a,m)J_{KKZ}(b)\tilde{s}^5. \tag{3.3.29}$$

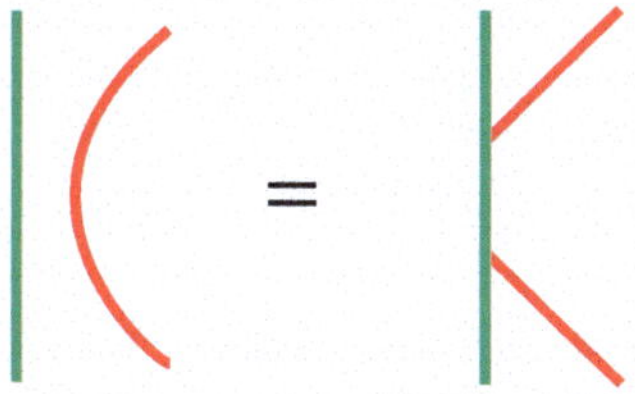

Fig. 3.16 Illustration of a $\mathbb{Z}_2$ symmetry defect partially absorbed into a duality defect. The green and red lines represent the duality and $\mathbb{Z}_2$ symmetry defects, respectively

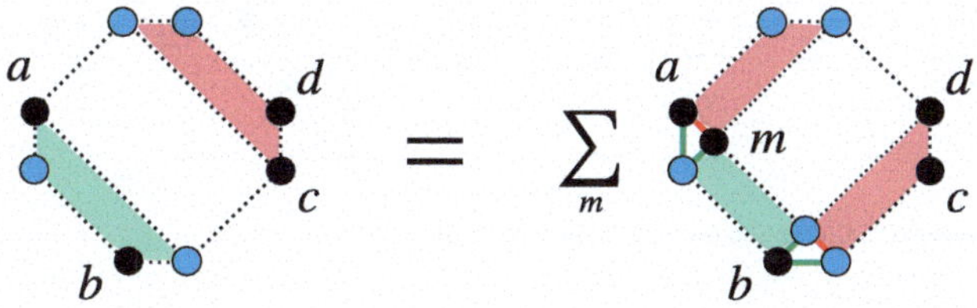

Fig. 3.17 Illustration of the a crossing relation of Fig. 3.16 on the lattice of the AMF approach. Black and blue dots represent active and inactive sites, respectively. The green quadrangle represents the building block of the duality symmetry defects. The red quadrangles represent the building blocks of the $\mathbb{Z}_2$ symmetry defects

The above two crossing relations mean that the duality defects can absorb $\mathbb{Z}_2$ symmetry defects. If we fuse those defects that are actually parallel in the theory with periodic boundary conditions, we see that the $\mathbb{Z}_2$ symmetry defect removes, as shown in Fig. 3.18. The first equality in Fig. 3.18 follows from Eq. (3.3.29). The middle equality follows from periodic boundary condition. The last equality follows

Fig. 3.18 $\mathbb{Z}_2$ symmetry defects absorb into duality defects and disappear. The square represents the spacetime in which the Ising model is defined. The top and bottom of the space satisfy periodic boundary conditions. The green and red lines represent the duality and $\mathbb{Z}_2$ symmetry defects, respectively

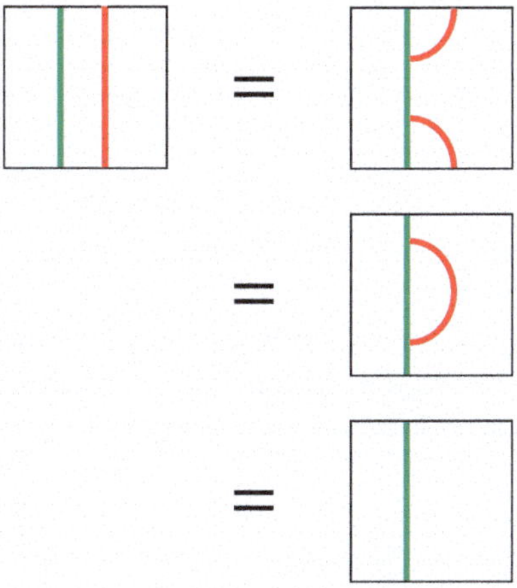

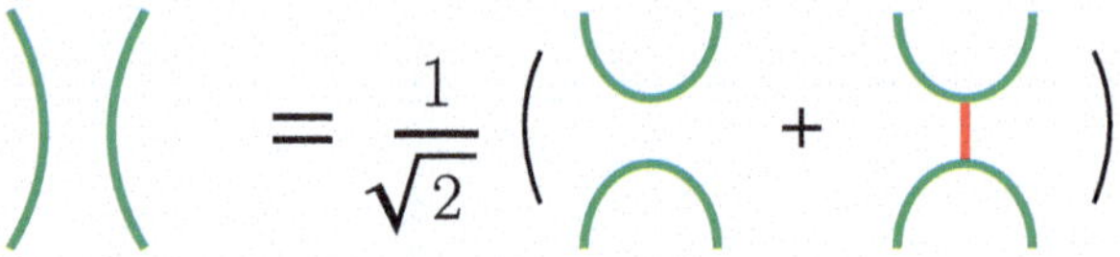

Fig. 3.19 Illustration of a crossing relation of duality defects. The green and red lines represent the duality and $\mathbb{Z}_2$ symmetry defects, respectively

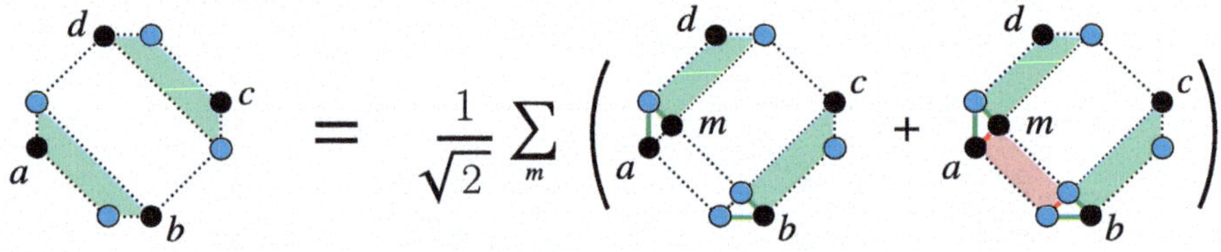

Fig. 3.20 Illustration of a crossing relation Fig. 3.19. Black and blue dots represent active and inactive sites, respectively. The green and red quadrangles represent the building blocks of the duality defects and the duality defects, respectively. The dotted quadrangle with the black dots a and m is the building block of the trivial defect

from Eq. (3.3.28). The formula for the interaction of two parallel defects is called the fusion rule. Let D denote the duality defect line and η denote the $\mathbb{Z}_2$ symmetry defect line. The fusion rule of these defects is expressed in

$$D \times \eta = D. \tag{3.3.30}$$

We consider the partial fusion of two parallel duality defect lines. When we partially fuse the duality defects with each other, we obtain a trivial defect and a $\mathbb{Z}_2$ symmetry defect. It is shown schematically in Fig. 3.14. This crossing relation can also be reproduced on the lattice as in Fig. 3.20. Figure 3.20 is represented by Eq. (3.3.31).

$$\begin{aligned} W(b.d)D(a,b)D(c,d)\tilde{s}^4 =& \frac{1}{\sqrt{2}}\{\sum_m W(c,m)D(b,c)D(m,d)Z(a,m)J_{KKZ}(b)J_{KKZ}(a,m)\tilde{s}^5 \\ &+\sum_m W(c,m)D(b,c)D(m,d)T(a,m)J_{KKT}(b)J_{KKT}(a,m)\tilde{s}^5\}. \end{aligned} \tag{3.3.31}$$

We consider the procedure shown in Fig. 3.21 on the Ising model with periodic boundary conditions. From this crossing relation, We obtain the fusion rule of duality defect lines D.

$$D \times D = 1 + \eta. \tag{3.3.32}$$

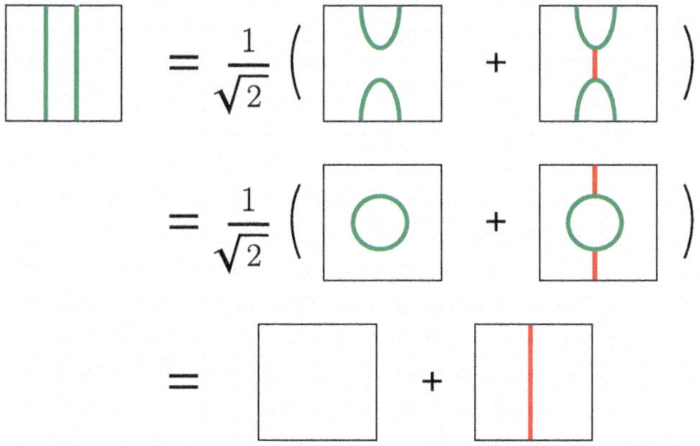

Fig. 3.21 The two duality defects fuse to form a trivial defect and a $\mathbb{Z}_2$ symmetry defect. The square represents the spacetime in which the Ising model is defined. The top and bottom of the space satisfy periodic boundary conditions. The green and red lines represent the duality and $\mathbb{Z}_2$ symmetry defects, respectively

The last equality follows from the fact that the quantum dimension of the duality defects is $\sqrt{2}$.

The duality defects are placed at the boundary of the Ising model and the gauged Ising model, which are associated with KW duality. Let us confirm this using the fusion rule. This is shown in Fig. 3.22 using defects. We prepare an Ising model with periodic boundary conditions. The theory supported in this region is defined as A. In the first equality, we place the duality defects on the space using the fact that the quantum dimension of the duality defects is $\sqrt{2}$. Let B be the theory supported in the region where theory A is supported and in the region connected across the duality defect. The second identity follows from the periodic boundary conditions. The third equality used the partial fusion of the duality defects. The last equality also uses the fact that the quantum dimension is $\sqrt{2}$. Finally, we find that theory A can be obtained from adding up all the topologically different configurations of the $\mathbb{Z}_2$ symmetry defects on theory B. This means that theory A can be obtained by gauging the $\mathbb{Z}_2$ symmetry of theory B.

We consider the crossing relation between two duality defects lines and $\mathbb{Z}_2$ symmetry defects lines connected by a single junction, such as the left side of Fig. 3.23, and defects with $\mathbb{Z}_2$ symmetry defects whose endpoints are located at those duality defects, such as the right side of Fig. 3.23. This crossing relation is expressed on the lattice as in Fig. 3.24 and in Eq. (3.3.33).

$$\begin{aligned} &W(c,d)D(a,d)D(b,c)J_{KKZ}(a,b)\tilde{s}^4 \\ &\quad = -\sum_{m_1,m_2} W(m_1,m_2)D(a,m_1)D(b,m_2)Z(m_1,d)Z(m_2,c)J_{KKZ}(a,b)J_{KKZ}(m_1,d)J_{KKZ}(m_2,c)\tilde{s}^5. \end{aligned} \tag{3.3.33}$$

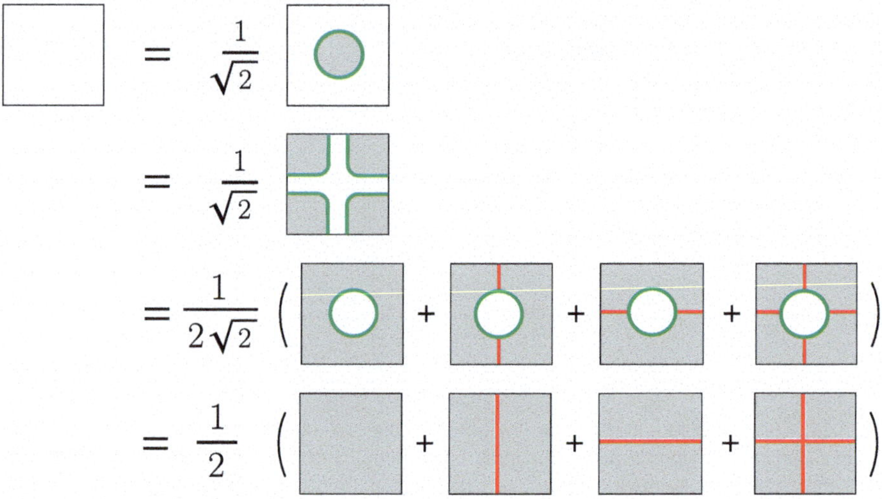

Fig. 3.22 The square represents a torus spacetime whose top and bottom, left and right satisfies periodic boundary conditions. Theories defined in the white region and theories defined in the gray region are KW dual. The white region theory is obtained by adding up the $\mathbb{Z}_2$ symmetry defects in all topologically different configurations on the gray region theory. This means that the white region theory is obtained by gauging the gray region theory with $\mathbb{Z}_2$ symmetry

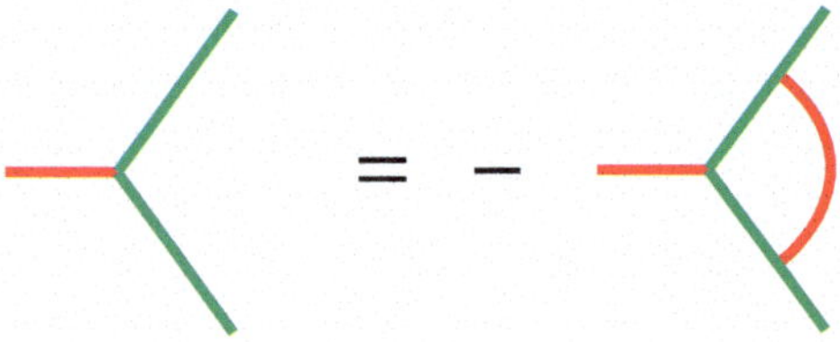

Fig. 3.23 Illustration of one of the crossing relation. The green and red lines represent the duality and $\mathbb{Z}_2$ symmetry defects, respectively

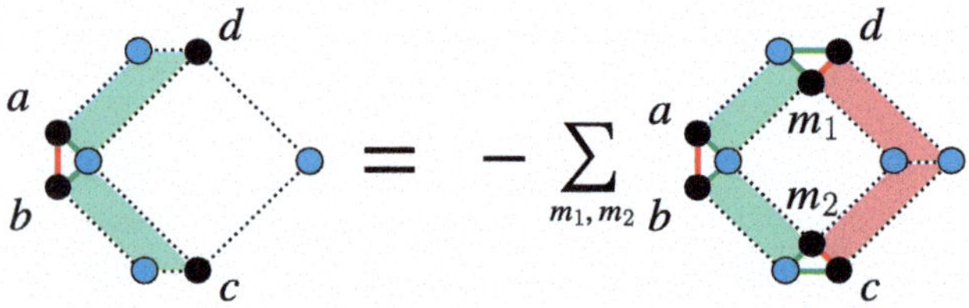

Fig. 3.24 Illustration of the a crossing relation of Fig. 3.23 on the lattice of the AMF approach. Black and blue dots represent active and inactive sites, respectively. The green quadrangle represents the building block of the duality symmetry defects. The red quadrangles represent the building blocks of the $\mathbb{Z}_2$ symmetry defects

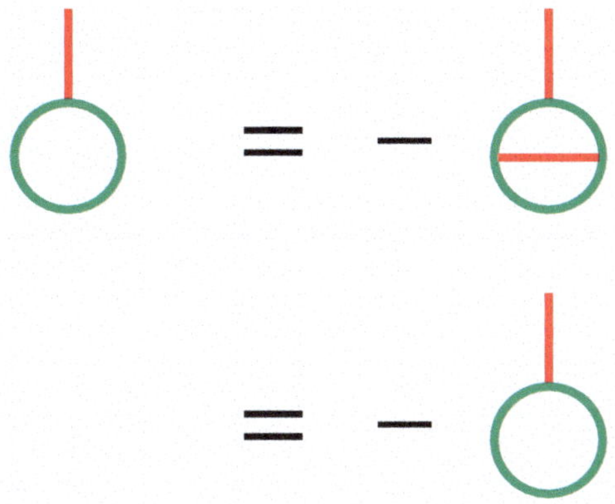

Fig. 3.25 Defects containing duality defect loops to which one $\mathbb{Z}_2$ symmetry defect is connected will have a value of zero. In the first equality, we use relation in Fig. 3.23 to generate a $\mathbb{Z}_2$ symmetry defect. In the second equality, we use relation in 3.14 to absorb the $\mathbb{Z}_2$ symmetry defect into the bottom of the duality defect loop

By crossing relation represented in Figs. 3.14, 3.23, we can show that the expected value of defects containing tadpole graph created by duality defects and $\mathbb{Z}_2$ symmetry defects is zero, as in Fig. 3.25. In the first equality in Fig. 3.25, we use crossing relation in Fig. 3.23 to generate $\mathbb{Z}_2$ symmetry defects inside the loop of duality defects. In the second equality, we used crossing relation in Fig. 3.14 to absorb $\mathbb{Z}_2$ defects inside the loop into the bottom part of the duality defects.

References

1. Koide, M., Nagoya, Y., Yamaguchi, S.: Non-invertible topological defects in 4- dimensional Z2 pure lattice gauge theory. PTEP **2022**(1), 013B03 (2022). arXiv: 2109.05992 [hep-th]
2. Aasen, D., Mong, R.S.K., Fendley, P.: Topological defects on the lattice I: the Ising model. J. Phys. A **49**(35), 354001 (2016). https://doi.org/10.1088/1751-8113/49/35/354001. arXiv:1601.07185 [cond-mat.stat-mech]

Chapter 4
KWW Defect in 4-Dimensional Lattice Gauge Theory

In this chapter, we explain KWW duality defects and 1-form $\mathbb{Z}_2$ symmetry defects in the 4-dimensional $\mathbb{Z}_2$ lattice gauge theory based on [1].

4.1 Kramers-Wannier-Wegner Duality

In this section, We confirm that this theory has the KW-like duality, it is Kramers-Wannier-Wegner duality.

We consider 4-dimensional square lattice system with periodic boundary condition. The number of site, link, plaquette, cube in this lattice are denoted by N_s, N_l, N_p, N_c, respectively. Then, their numbers are related to the following,

$$N_l = N_c = 4N_s, \tag{4.1.1}$$
$$N_p = 6N_s. \tag{4.1.2}$$

We denote the link consisting with sites i_1, i_2 by $\langle i_1 i_2 \rangle$, the cube with plaquettes p_1, p_2, p_3, p_4, p_5, p_6 on boundary by $\langle p_1 p_2 p_3 p_4 p_5 p_6 \rangle$ respectively. And we denote the plaquette with links l_1, l_2 l_3, l_4 on boundary by $\langle l_1 l_2 l_3 l_4 \rangle$. We assign the link variable $a_l = 0, 1$ to the link l. The partition function of this model is

$$Z_{\mathcal{T}} = \sum_{\{a\}} \exp \left\{ \sum_{\substack{\text{plaquette} \\ \langle l_1 l_2 l_3 l_4 \rangle}} K(-1)^{a_{l_1}+a_{l_2}+a_{l_3}+a_{l_4}} \right\}. \tag{4.1.3}$$

As studied in Sect. 2.2.1, this theory has 1-form $\mathbb{Z}_2^{(1)}$ global symmetry. Let us define the partition function of the 1-form $\mathbb{Z}_2^{(1)}$ gauged lattice gauge theory $Z_{\mathcal{T}/\mathbb{Z}_2^{(1)}}$.

M. Koide, *Non-invertible Symmetry in 4-Dimensional Z2 Lattice Gauge Theory*, Springer Theses, https://doi.org/10.1007/978-981-96-2272-6_4

In Sect. 3.1, we obtained the 0-form $\mathbb{Z}_2$ gauged Ising model by adding the background 1-form flat gauge field b_l and summing up all configuration of b_l. As an analogy to this, by adding the background 2-form flat gauge field b_p on plaquette and summing up all configuration of it, we obtain the 1-form $\mathbb{Z}_2^{(1)}$ gauged lattice gauge theory $Z_{T/\mathbb{Z}_2^{(1)}}$. First, we assign the 2-form gauge field $b_p = 0, 1$ on the plaquette p. This 2-form gauge field b_p couple with link variable a_l on a plaquette $p = \langle l_1 l_2 l_3 l_4 \rangle$ as follow,

$$\exp\left\{K(-1)^{a_{l_1}+a_{l_2}+a_{l_3}+a_{l_4}+b_p}\right\}. \tag{4.1.4}$$

To make the 2-form gauge field b_p flat, we also introduce following action on the cube $\gamma = \langle p_1 p_2 p_3 p_4 p_5 p_6 \rangle$,

$$\frac{1}{2}\sum_{c_\gamma=0,1} \exp\left\{i\pi c_\gamma (b_{p_1} + b_{p_2} + b_{p_3} + b_{p_4} + b_{p_5} + b_{p_6})\right\}. \tag{4.1.5}$$

Then, these action is invariant by 1-form gauge transformation,

$$a_l \to a_l + \lambda_l \quad (\lambda_l = 0, 1), \tag{4.1.6}$$

$$\begin{aligned} b_{l_1 l_2 l_3 l_4} &\to b_{l_1 l_2 l_3 l_4} + \delta\lambda_{l_1 l_2 l_3 l_4} \\ &\equiv b_{l_1 l_2 l_3 l_4} + \lambda_{l_1} + \lambda_{l_2} + \lambda_{l_3} + \lambda_{l_4}. \end{aligned} \tag{4.1.7}$$

Pay attention to the calculation of the gauge volume. the parameter $\delta\lambda_{l_1 l_2 l_3 l_4}$ is invariant by the 0-form gauge transformation,

$$\lambda_{l=\langle ij \rangle} \to \lambda_{l=\langle ij \rangle} + \mu_i + \mu_j \quad (\mu_i,\ \mu_j = 0, 1,\ i,\ j : \text{site}), \tag{4.1.8}$$

$$\delta\lambda_{l_1 l_2 l_3 l_4} \to \delta\lambda'_{l_1 l_2 l_3 l_4} = \delta\lambda_{l_1 l_2 l_3 l_4}. \tag{4.1.9}$$

So, if we count 2 gauge degrees of freedom for each link, that includes 2 overcounts for each site. Therefore the correct gauge volume is

$$\frac{2^{N_l}}{2^{N_s}}. \tag{4.1.10}$$

The partition function of the 1-form $\mathbb{Z}_2^{(1)}$ gauged lattice gauge theory $Z_{T/\mathbb{Z}_2^{(1)}}$ is

$$Z_{\mathcal{T}/\mathbb{Z}_2^{(1)}} = \frac{2^{N_s}}{2^{N_l}2^{N_c}} \sum_{\{a\}\{b\}\{c\}} \exp\left\{ \sum_{\substack{\text{plaquette} \\ p=\langle l_1 l_2 l_3 l_4\rangle}} K(-1)^{a_{l_1}+a_{l_2}+a_{l_3}+a_{l_4}+b_p} \right\}$$
$$\times \exp\left\{ i\pi \sum_{\substack{\text{cube} \\ \gamma=\langle p_1 p_2 p_3 p_4 p_5 p_6\rangle}} c_\gamma (b_{p_1}+b_{p_2}+b_{p_3}+b_{p_4}+b_{p_5}+b_{p_6}) \right\}. \tag{4.1.11}$$

Since the configurations connected by the gauge transformation are physically the same, we perform a gauge fixing $a_l = 0$ and sum up $\{a\}$.

$$Z_{\mathcal{T}/\mathbb{Z}_2^{(1)}} = \frac{2^{N_s}}{2^{N_c}} \sum_{\{b\}\{c\}} \exp\left\{ \sum_{\substack{\text{plaquette} \\ p=\langle l_1 l_2 l_3 l_4\rangle}} K(-1)^{b_p} \right\}$$
$$\times \exp\left\{ i\pi \sum_{\substack{\text{cube} \\ \gamma=\langle p_1 p_2 p_3 p_4 p_5 p_6\rangle}} c_\gamma (b_{p_1}+b_{p_2}+b_{p_3}+b_{p_4}+b_{p_5}+b_{p_6}) \right\}. \tag{4.1.12}$$

The Kramers-Wannier-Wegner duality is a relationship between $Z_{\mathcal{T}}(K)$ and $Z_{\mathcal{T}/\mathbb{Z}_2^{(1)}}(K)$. Its exact relationship is the following,

$$\frac{1}{(\sinh 2K)^{3N_s/2}} Z_{\mathcal{T}}(K) = \frac{1}{(\sinh 2\tilde{K})^{3N_s/2}} Z_{\mathcal{T}/\mathbb{Z}_2^{(1)}}(\tilde{K}), \tag{4.1.13}$$

$$\sinh 2K \sinh 2\tilde{K} = 1. \tag{4.1.14}$$

As in the case of the Ising model, rewrite the original theory to investigate its correspondence with the gauged theory. By using the Eq. (3.1.15), we deform the partition function $Z_{\mathcal{T}}$,

$$Z_{\mathcal{T}}(K) = \frac{1}{(2\sinh 2\tilde{K})^{N_p/2}} \sum_{\{a\}\{b\}} \prod_{\substack{\text{plaquette} \\ p=\langle l_1 l_2 l_3 l_4\rangle}} \exp\left\{\tilde{K}(-1)^{b_p}\right\}$$
$$\times \exp\left\{i\pi(a_{l_1}+a_{l_2}+a_{l_3}+a_{l_4})b_p\right\}. \tag{4.1.15}$$

The term $(a_{l_1}+a_{l_2}+a_{l_3}+a_{l_4})b_p$ focuses on the plaquette $p = \langle l_1 l_2 l_3 l_4\rangle$ and describes the interaction between the link variables and the plaquette variables. Rewrite this term in a form that focuses on each link. In a 4-dimensional square lattice, each link is located on the boundary of 6 different plaquettes. We denote the plaquette with link l on the boundary by $p(l)$. Then the partition function is rewritten as follows.

$$Z_{\mathcal{T}}(K) = \frac{1}{(2\sinh 2\tilde{K})^{N_p/2}} \sum_{\{a\}\{b\}} \exp\left\{\tilde{K} \sum_{\substack{\text{plaquette}\\ \text{p}}} (-1)^{b_p}\right\}$$
$$\times \exp\left\{i\pi \sum_{\substack{\text{link}\\ l}} a_l (b_{p_1(l)} + b_{p_2(l)} + b_{p_3(l)} + b_{p_4(l)} + b_{p_5(l)} + b_{p_6(l)})\right\}. \tag{4.1.16}$$

We introduce a dual lattice by placing dual sites in the center of the hypercube of the lattice and dual links between re-adjacent dual sites. By periodic boundary conditions, this dual lattice has the number of sites, the number of links, the number of plaquettes, and the number of cubes equal to that of the original lattice. At this time, the links and plaquettes of the original lattice are orthogonal to the cubes and plaquettes of the dual lattice and at their centers, respectively We denote a label of dual cube orthogonal to a original link l by $\tilde{l}$, and a label of dual plaquette orthogonal to a original plaquette p by $\tilde{p}$. And we define the dual cube variable $\tilde{c}_{\tilde{l}}$ on the dual cube $\tilde{l}$ by the original link variable $a_l = 0, 1$ and the dual plaquette variable $\tilde{b}_{\tilde{p}}$ on dual plaquette $\tilde{p}$ by the original plaquette variable $b_p = 0, 1$. If an original link l is placed on a boundary of an original plaquette $p = p(l)$, the dual cube $\tilde{l}$ have the dual plaquette on the boundary. If a dual plaquette $\tilde{p}$ placed on a boundary of a tilde cube $\tilde{l}$, we denote the plaquette by $\tilde{p} = \tilde{p}(\tilde{l})$. In term of the dual lattice, the partition function is written by

$$Z_{\mathcal{T}}(K) = \frac{1}{(2\sinh 2\tilde{K})^{N_p/2}} \sum_{\{\tilde{a}\}\{\tilde{b}\}} \exp\left\{\tilde{K} \sum_{\substack{\text{plaquette}\\ \tilde{p}}} (-1)^{\tilde{b}_{\tilde{p}}}\right\}$$
$$\times \exp\left\{i\pi \sum_{\substack{\text{cube}\\ \tilde{l}}} \tilde{a}_{\tilde{l}} (\tilde{b}_{\tilde{p}_1(\tilde{l})} + \tilde{b}_{\tilde{p}_2(\tilde{l})} + \tilde{b}_{\tilde{p}_2(\tilde{l})} + \tilde{b}_{\tilde{p}_4(\tilde{l})} + \tilde{b}_{\tilde{p}_5(\tilde{l})} + \tilde{b}_{\tilde{p}_6(\tilde{l})})\right\}. \tag{4.1.17}$$

Since $N_p/2 = 3N_s = N_l - N_s$ follows from Eq. (4.1.2), we have proved KWW duality (4.1.13).

Similar to the 2-dimensional Ising model, KWW duality is self-duality when $K = K_c = \tilde{K} = -1/2 \log -1 + \sqrt{2}$.

$$Z_{\mathcal{T}}(K_c) = Z_{\mathcal{T}/\mathbb{Z}_2^{(1)}}(\tilde{K}_c), \tag{4.1.18}$$

In the following, the parameters are fixed to $K = K_c$.

4.2 4-Dimensional $\mathbb{Z}_2$ Lattice Gauge Theory

In this chapter, we describe our approach to the 4-dimensional $\mathbb{Z}_2$ lattice gauge theory. To prepare for future discussions on duality defect, we introduce two types of lattice, as illustrated in Fig. 4.1. Our 4-dimensional approach closely resembles the one used in 2-dimensions in [2].

To provide a more detailed description of the lattices, we introduce a 4-dimensional coordinate system, denoted as (x_1, x_2, x_3, x_4) in $\mathbb{R}^4$. Points within Λ have coordinates consisting of even integers, defined as $\Lambda := \{(x_1, x_2, x_3, x_4)|x_1, x_2, x_3, x_4 \in 2\mathbb{Z}\}$. Conversely, for $\hat{\Lambda}$, the coordinates are formed by odd integers, expressed as $\hat{\Lambda} := \{(x_1, x_2, x_3, x_4)|x_1, x_2, x_3, x_4 \in 2\mathbb{Z} + 1\}$. We refer to the line connecting the closest pair of points in each lattice as a link. We also use the term plaquette for the smallest square formed by these links. However, it is important to note that we do not classify the line connecting a point on Λ with a point on $\hat{\Lambda}$ as a link.

We assign the link variables $U_m = \pm 1$ to the links in Λ, while we do not assign link variables to the links in $\hat{\Lambda}$. For this reason, we refer to the lattice Λ, to which the link variables are assigned, as the active lattice. Consequently, we label the sites, links, and plaquettes on the lattice Λ as active sites, active links, and active plaquettes respectively. On the other hand, the lattice $\hat{\Lambda}$, where no link variables are assigned, is referred to as the inactive lattice. In this context, we label the sites, links, and plaquettes on this lattice as inactive sites, inactive links, and inactive plaquettes.

We can consider the basic unit of this lattice system as a 16-cell (for more details, see [3, 4]). As depicted in Fig. 4.2, this 16-cell consists of an active plaquette and an inactive plaquette that share a common center. Within the 16-cell, there are 16 tetrahedrons, each of which contains one active link and one inactive link. The surface of the 16-cell is homeomorphic to S^3.

The active placket and 16 cells correspond one-to-one. For example, we consider an active plaquette p composed of four points at coordinates $(0, 0, 0, 0)$, $(0, 0, 0, 2)$, $(0, 0, 2, 0)$, $(0, 0, 2, 2)$ in Λ. The center point of this plaquette is $(0, 0, 1, 1)$. Then

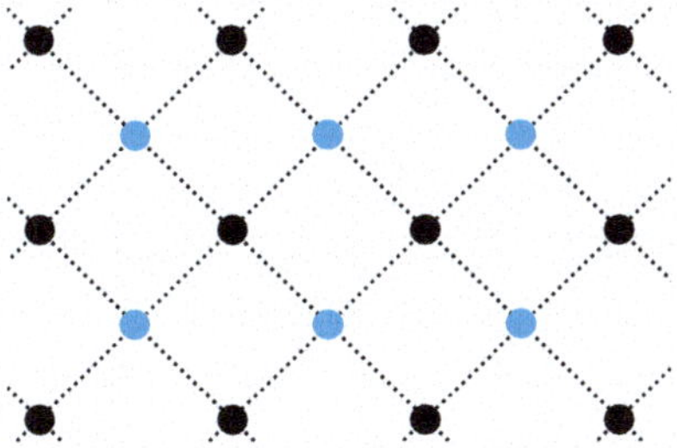

Fig. 4.1 A schematic illustration of lattices. Although these lattices are depicted as 2-dimensional in this figure, the actual lattices treated in this paper are 4-dimensional ones. The black lattice consisting of the black dots represents the lattice Λ, while the blue lattice consisting of the blue dots represents the lattice $\hat{\Lambda}$. These lattices are mutually dual. Reprinted from [1], Copyright 2021, The Author(s) licensed under CC BY 4.0

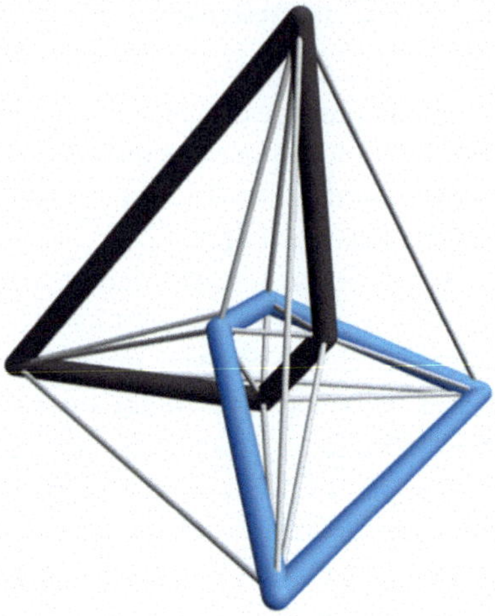

Fig. 4.2 A stereographic projection of a 16-cell into 3 dimensions. The black plaquette represents an active plaquette, and the blue plaquette represents an inactive plaquette. The 16-cell consists of 16 tetrahedrons, each containing one active link and one inactive link. Reprinted from [1], Copyright 2021, The Author(s) licensed under CC BY 4.0

the inactive plaquette $\tilde{p}$ composed of four points at coordinates $(\pm 1, \pm 1, 1, 1)$ in $\hat{\Lambda}$ shares the same center as p. There is a 16-cell that includes both p, $\tilde{p}$, and these eight points.

This one-to-one correspondence between the 16 cells and the active plaquette makes assigning Boltzmann weights to the active plaquette the same as assigning Boltzmann weights to the 16 cells. Let $a_i = 0, 1$ for $(i = 1, 2, 3, 4)$ be the link variables for the four active links within a 16-cell. We define the Boltzmann weight to this 16-cell as

$$W(a_1, a_2, a_3, a_4) = \exp(K(-1)^{(a_1+a_2+a_3+a_4)}). \tag{4.2.1}$$

This definition is same as the Boltzmann weight in the $\mathbb{Z}_2$ gauge theory.

To define the duality defect later, we introduce the weights for active links, active sites, inactive links, and inactive sites as s, l, $\bar{s}$, and $\bar{l}$, respectively. These specific value of the weights is determined in Sect. 4.3.1.

We define the partition function of this theory as

$$Z = \sum_{\{a\}} \left(\prod_{\substack{\text{active} \\ \text{sites}}} s \right) \left(\prod_{\substack{\text{active} \\ \text{links}}} l \right) \left(\prod_{\substack{\text{inactive} \\ \text{sites}}} \bar{s} \right) \left(\prod_{\substack{\text{inactive} \\ \text{links}}} \bar{l} \right) \prod_{i \in C} W(a_{j_1(i)}, a_{j_2(i)}, a_{j_3(i)}, a_{j_4(i)}), \tag{4.2.2}$$

where C is the set of all 16-cells, and $j_1(i)$, $j_2(i)$, $j_3(i)$, and $j_4(i)$ are the four active links in the 16-cell i. a_j is the link variable assigned to the active link j. Besides the constant normalization, it is the same as the partition function (4.1.3).

4.3 Topological Defect

In this section, we explain topological defects and junctions among them in the 4-dimensional $\mathbb{Z}_2$ lattice gauge theory following AMF approach [2, 5]. We introduce defects and junctions into our lattice system. We impose the defect commutation relations of topological defects and determine their weights so that they hold. We derive crossing relations among them. We calculate the expectation values of some configurations of defects by using these crossing relations. In particular, we find that the defect associated to the KWW duality is non-invertible.

4.3.1 Duality Defect

In this subsection, we construct duality defects in the 4-dimensional $\mathbb{Z}_2$ pure gauge theory. Duality defects are 3-dimensional operators. It is then placed on the boundary of the dual theories associated with the KWW duality. The active and inactive lattices are swapped across the duality defects, like the KW defects in a 2-dimensional Ising model [2].

The 3-dimensional unit cell on the tessellation by regular 16-cells is a regular tetrahedron located on the surface of those 16 cells. Therefore, we employ these tetrahedrons as the building blocks of the 3-dimensional surface on which a duality defect is supported. The active lattice and the inactive lattice are swapped across the duality defect. In order to implement this property, it is convenient to double each tetrahedron on which the duality defect is supported, and swap the active link and the inactive link (see Fig. 4.3). As a result, the building block of duality defects is a tetrahedral prism (see Fig. 4.4). This is analogous to constructing the KW duality defect in the 2-dimensional Ising model by doubling link and swap the active site and the inactive site in [2].

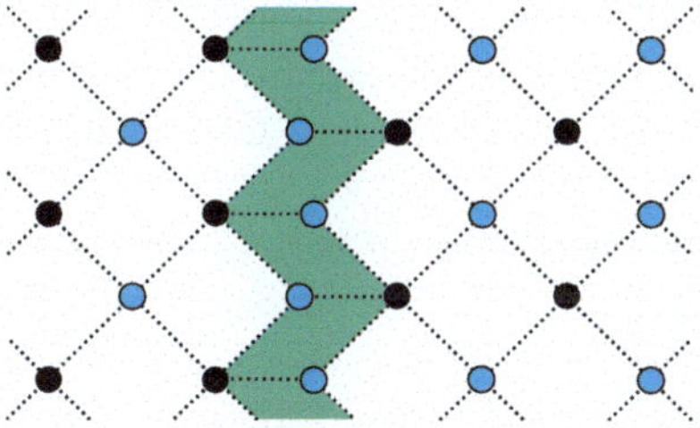

Fig. 4.3 Schematic illustration of the duality defect. The black dots represent the active lattice and the blue dots represent the inactive lattice. A duality defect is located at the boundary between the two regions. The active lattice (black dots) and the inactive lattice (blue dots) are swapped across the duality defect. A unit cell of the duality defect is a tetrahedral prism depicted as a green parallelogram in this figure. Reprinted from [1], Copyright 2021, The Author(s) licensed under CC BY 4.0

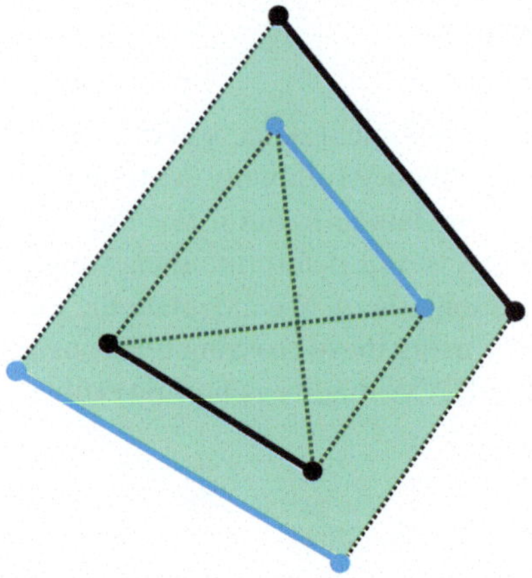

Fig. 4.4 The building block of duality defects in the 4-dimensional $\mathbb{Z}_2$ lattice gauge theory. The 3-dimensional surface is composed of tetrahedrons each of which includes an active link and an inactive link. A tetrahedron on which a duality defect is supported is doubled and becomes a tetrahedral prism. An inactive link is put on the edge of a tetrahedron in this tetrahedral prism associated to the active link on the other tetrahedron and vice versa. Reprinted from [1],

The KWW duality transformation just changes the description of the theory, but not change the observables. Therefore we expect that topological duality defects exist like the KW defects in a 2-dimensional Ising model. In order to construct such topological defects, we impose the defect commutation relations so that the duality defect is topological, and find a solution.

Since a 16-cell is a basic unit of our lattice system, we focus on a single 16-cell and consider defect commutation relations (see Fig. 4.5). We consider a configuration of a duality defect A. Some of the 16 tetrahedrons in the focused 16-cell are filled by this defect A and the others are not. We also consider a deformed configuration B in which the tetrahedrons on the focused 16-cell which are not filled by A is filled and vice versa. So, A and B differ only on the focused 16-cell. For the duality defect is topological, A and B must have the same weight if A and B have the same topology. Such an equality between two different duality defect configurations is called a defect commutation relation.

In the defect commutation relation, it is necessary to consider coordination in which the topology of the defect does not change before and after the deformation. Then the topology of A and B are the same without any ambiguity if and only if these configurations on the focused 16-cell satisfy the following conditions.

- The tetrahedrons filled in the 16-cell for both A and B are not empty.
- Both A and B configurations, restricted to the surface of the focused 16-cell, are simply connected.
- There is no ambiguity in the above condition. In other words, there is no connection in which the duality defects are connected only by sites or links.

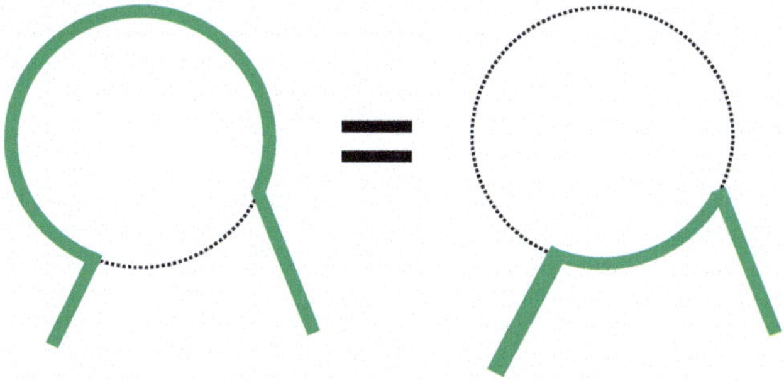

Fig. 4.5 A schematic illustration of a defect commutation relation. The circle represents a 16-cell and a green line represents a duality defect. The defect commutation relation implies that the value of the duality defect remains the same even if it is deformed without changing the topology. Reprinted from [1], Copyright 2021, The Author(s) licensed under CC BY 4.0

We provide a more detailed explanation of the defect commutation relations. Each building block of a duality defect contains two active links. Let us take a building block and we assign the link variables $a, \tilde{a} = 0, 1$ to these two active links. Then we can assign a weight $D(a, \tilde{a})$ to this building block. In Sect. 4.2, we defined the weights for both the active lattice and the inactive lattice components. Specifically, we denote the weight of an active link as l, an active site as s, an inactive link as $\bar{l}$, and an inactive site as $\bar{l}$.

We explain in more detail about the notations. Let's choose a 16 cell with no duality defects on the surface with our system. We assign the labels $m = 1, 2, 3, 4$ to the active links and the labels $\tilde{n} = \tilde{1}, \tilde{2}, \tilde{3}, \tilde{4}$ to the inactive links. Then we denote the set of four active links in the 16-cell as $M = \{1, 2, 3, 4\}$ and the set of four inactive links as $\tilde{N} = \{\tilde{1}, \tilde{2}, \tilde{3}, \tilde{4}\}$. A pair of an active link and an inactive link, denoted as $(m, \tilde{n})$, can be employed to identify a tetrahedron that holds these links within the 16-cell. Let U be the set of all tetrahedrons

$$U = \{(m, \tilde{n}) | m = 1, 2, 3, 4, \ \tilde{n} = \tilde{1}, \tilde{2}, \tilde{3}, \tilde{4}\}. \tag{4.3.1}$$

The weight of this 16-cell is the Boltzmann weight $W(a_1, a_2, a_3, a_4)$ in Eq. (4.2.1). Here, a_m is the link variable assigned to the active link m.

Next, we consider a 16-cell with duality defects on its surface in the configuration A. Let $I \subset U$ be the set of tetrahedrons filled by the duality defects in the configuration A. Since the duality defect is defined by doubling the tetrahedron, we need to take into account the active links that arise from this doubling, in addition to the set of active links M contained within this 16-cell. These additional active links are the counterparts of the inactive links included in tetrahedrons contained in the set I. Therefore We can use $\tilde{n}$, which originally serves as the label for an inactive link in the 16-cell, as the label for such additional active links. Therefore, the set of all such additional active links is $\tilde{E} = \{\tilde{n} | (m, \tilde{n}) \in I\}$. Also, let $\tilde{a}_{\tilde{n}}$ be the link variable of the additional active link $\tilde{n}$. By using these notations, we can denote the weight of each building block of this duality defect located on the surface of the 16-cell as

$$D(a_m, \tilde{a}_{\tilde{n}}), \quad (m, \tilde{n}) \in I. \tag{4.3.2}$$

Then the total weight of the 16-cell and the duality defect located on the 16-cell in the configuration A is written as

$$W(a_1, a_2, a_3, a_4) \prod_{(m,\tilde{n}) \in I} D(a_m, \tilde{a}_{\tilde{n}}). \tag{4.3.3}$$

The defect commutation relations state that the weight of the configuration, A, before changing as shown in Fig. 4.5, with the defect on I, equals the weight of the configuration, B, after changing, with the defect on the opposite area $\bar{I} = U \setminus I$. So, it's important to also look at the configuration B. It's key to note that in B, the 16-cell and its active connections are part of the area opposite the defect, where the active lattice and inactive lattice switch places. To make it easier to compare with the weights in configuration A, we can refer to the active links in the 16-cell post-deformation as $\tilde{n} = \tilde{1}, \tilde{2}, \tilde{3}, \tilde{4}$, while the inactive links in the 16-cell are called $m = 1, 2, 3, 4$. The variable for link $\tilde{n}$ is marked as $\tilde{a}_{\tilde{n}}$ because it connects to the system's other part just like $\tilde{a}_{\tilde{n}}$ in configuration A, if it exists. Using this label system, the Boltzmann weight for the 16-cell is shown as $W(\tilde{a}_{\tilde{1}}, \tilde{a}_{\tilde{2}}, \tilde{a}_{\tilde{3}}, \tilde{a}_{\tilde{4}})$. The combined weight of the 16-cell and duality defects on the 16-cell in configuration B is noted as

$$W(\tilde{a}_{\tilde{1}}, \tilde{a}_{\tilde{2}}, \tilde{a}_{\tilde{3}}, \tilde{a}_{\tilde{4}}) \prod_{(m,\tilde{n}) \in \bar{I}} D(a_m, \tilde{a}_{\tilde{n}}). \tag{4.3.4}$$

We also take into account $E = \{m | (m, \tilde{n}) \in \bar{I}\}$. This represents the collection of inactive links m adjacent to the duality defect situated on the tetrahedrons in $\bar{I}$. Every one of these inactive links is doubled by the duality defect, leading to a corresponding extra active link. The link variable a_m for this active link m connects to the rest of the system in a similar manner as the link variable a_m in the setup A. As a result, they are considered equivalent.

We set specific defect commutation relations as requirements for $D(a, \tilde{a}), l, s, \bar{l}, \bar{s},\ K$.

$$\sum_{M \setminus E} W(a_1, a_2, a_3, a_4) s^{\alpha_1} l^{\beta_1} \bar{s}^{\tilde{\alpha_1}} \bar{l}^{\tilde{\beta_1}} \prod_{(m,\tilde{n}) \in I} D(a_m, \tilde{a}_{\tilde{n}}) = \sum_{\tilde{N} \setminus \tilde{E}} W(\tilde{a}_{\tilde{1}}, \tilde{a}_{\tilde{2}}, \tilde{a}_{\tilde{3}}, \tilde{a}_{\tilde{4}}) s^{\alpha_2} l^{\beta_2} \bar{s}^{\tilde{\alpha_2}} \bar{l}^{\tilde{\beta_2}} \prod_{(m,\tilde{n}) \in \bar{I}} D(a_m, \tilde{a}_{\tilde{n}}). \tag{4.3.5}$$

In this context, the counts of active sites, active links, inactive sites, and inactive links within the configuration A are represented by $\alpha_1, \beta_1, \tilde{\alpha_1}$, and $\tilde{\beta_1}$ respectively. Similarly, for configuration B, the numbers of active sites, active links, inactive sites, and inactive links are indicated by $\alpha_2, \beta_2, \tilde{\alpha_2}$, and $\tilde{\beta_2}$ respectively. Furthermore, the sums of $\sum_{M \setminus E}, \sum_{\tilde{N} \setminus \tilde{E}}$ are defined by

$$\sum_{M\setminus E} := \prod_{m\in M\setminus E}\sum_{a_m=0,1}, \quad \sum_{\tilde{N}\setminus\tilde{E}} := \prod_{\tilde{n}\in\tilde{N}\setminus\tilde{E}}\sum_{\tilde{a}_{\tilde{n}}=0,1}. \tag{4.3.6}$$

In Eq. (4.3.5), on the left side we add up the link variables for the links in $M \setminus E$, and on the right side, the addition is for the link variables of the links in $\tilde{N} \setminus \tilde{E}$. Here's more detail about this addition. Initially, both sides of the defect commutation relations represent the input to the partition function or correlation functions with the defects from the 16-cell and its components on its surface. The defect commutation relations must be fulfilled for any link variables in the active links found in other 16-cells, where a different operator might be added. On the other side, the active links in $M \setminus E$, $\tilde{N} \setminus \tilde{E}$ are not part of other 16-cells. These represent the degrees of freedom that emerge and vanish before and after the deformation. Thus, sums derived from the partition function are applied to the link variables of such links.

We find the values $D(a, \tilde{a}), l, s, \bar{l}, \bar{s}, K$ by resolving these defect commutation relations. In a solution that makes physical sense, these values fulfill

$$D(a,\tilde{a}) \neq 0, \quad l, s, \bar{l}, \bar{s} > 0, \quad K \in \mathbb{R}, \quad K \neq 0. \tag{4.3.7}$$

There is a single solution that makes physical sense, except for the sign of $D(a, \tilde{a})$. The solution is[1]

$D(a,\tilde{a}) =$ 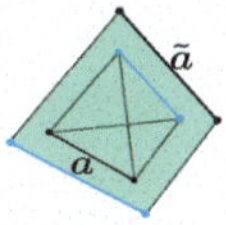

$= (-1)^{a\tilde{a}},$ (4.3.8)

$l =$  $= \frac{1}{\sqrt{2}},$ $s =$  $= \frac{1}{\sqrt{2}},$ (4.3.9)

$\bar{l} =$  $= 1,$ $\bar{s} =$  $= 1,$ (4.3.10)

$$K = K_c = -\frac{1}{2}\log(-1+\sqrt{2}), \qquad W(a_1, a_2, a_3, a_4) = \exp(K(-1)^{(a_1+a_2+a_3+a_4)}). \tag{4.3.11}$$

We demonstrate the gauge invariance of our duality defects. A single building block is not gauge invariant by itself. To understand this, let's consider the gauge transformation at a site S within a building block B. Define L as the active link at S that is part of B. Also, consider L' as another active link within B. Variables a and $\tilde{a}$ represent the link variables for L and L', respectively. Then the weight of this building block B is transformed by this gauge transformation as

[1] We have used Mathematica to find the solution. We have learned this method from Kantaro Ohmori's lecture "Categorical symmetry in 1 + 1 dimensions" (in Japanese) in CREST online workshop "Theoretical studies of topological phases of matter." We would like to thank him for this excellent lecture.

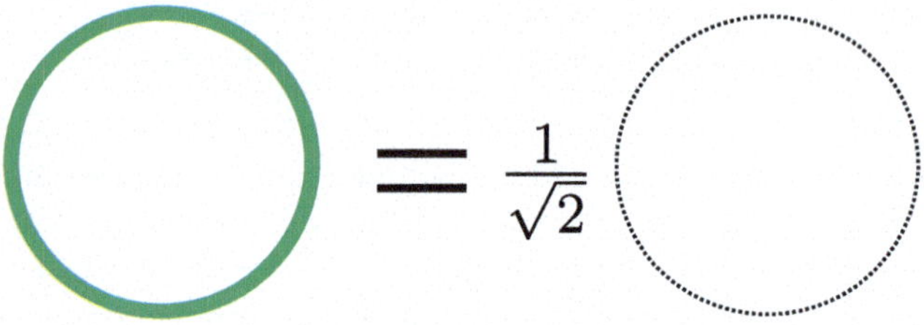

Fig. 4.6 A schematic illustration of Eq. (4.3.14). The circles represent the 16-cell and the green line represents the duality defect. Reprinted from [1], Copyright 2021, The Author(s) licensed under CC BY 4.0

$$D(a, \tilde{a}) \quad \rightarrow \quad D(1-a, \tilde{a}) = (-1)^{(1-a)\tilde{a}}. \tag{4.3.12}$$

So, it is not gauge invariant. The entire duality defect is gauge invariant, as demonstrated below. Because the entire duality defect lacks a boundary, there exists a unique building block that includes S and L' but excludes L; this building block is referred to as B'. Further, let b represent the link variable of the active link that includes S within B'. Then the weights of these building blocks B, B' are transformed by the gauge transformation at the active site S as

$$D(a, \tilde{a})D(b, \tilde{a}) \quad \rightarrow \quad D(1-a, \tilde{a})D(1-b, \tilde{a}) = (-1)^{(1-a)\tilde{a}}(-1)^{(1-b)\tilde{a}} = (-1)^{a\tilde{a}}(-1)^{b\tilde{a}} = D(a, \tilde{a})D(b, \tilde{a}). \tag{4.3.13}$$

Therefore, the pair of building blocks B, B' remains unchanged by this gauge transformation. Building blocks that include the site S are grouped into such pairs, and so remain unchanged. Those building blocks not including S are obviously unchanged. Hence, we can deduce that the entire duality defect is unaffected by gauge transformation.

We now demonstrate that the duality defects created here are non-invertible. To do this, we consider the situation where the set of filled tetrahedrons in the 16-cell for the configuration B is empty as depicted in Fig. 4.6. In the configuration A, the duality defect is present on all tetrahedrons of the 16-cell, creating a closed manifold S^3. In this case, the transformation changes the topology of the defect from S^3 to empty, hence, Eq. (4.3.5) is not required. Instead, one observes that the equation below is fulfilled by inserting our solution (4.3.8)–(4.3.11).

$$\sum_{a_1, a_2, a_3, a_4 = 0,1} W(a_1, a_2, a_3, a_4) s^8 l^8 \bar{s}^8 \bar{l}^8 \prod_{(m, \bar{n}) \in U} D(a_m, \tilde{a}_{\bar{n}}) = \frac{1}{\sqrt{2}} W(\tilde{a}_{\bar{1}}, \tilde{a}_{\bar{2}}, \tilde{a}_{\bar{3}}, \tilde{a}_{\bar{4}}) s^4 l^4 \bar{s}^4 \bar{l}^4. \tag{4.3.14}$$

For a symmetry defect placed on a closed manifold without any operator insertion inside, the weight is identical with the empty configuration. However, for our duality defect, they are not identical, but their ratio is $1/\sqrt{2}$. Therefore, we can conclude that our duality defects are non-invertible.

4.3.2 $\mathbb{Z}_2$ 1-Form Symmetry Defects

In this subsection, we construct 1-form $\mathbb{Z}_2$ symmetry defects. As explained in Sect. 2.2.1, these defects are defined on 2-dimensional surfaces. The charge operator is the Wilson loop. The symmetry defect acts as a link to the Wilson loop, flipping its sign.

Since $\mathbb{Z}_2$ symmetry defects are not codimension 1, their construction is somewhat different compared to the duality defects discussed in Sect. 4.3.1, as well as the defects constructed in [2, 5]. We examine a 2-dimensional closed surface made of triangles, each formed by an inactive link and the midpoint of a neighboring active link. We focus on a $\mathbb{Z}_2$ symmetry defect supported by this surface. To explain these $\mathbb{Z}_2$ symmetry defects, we deform the lattice as described below. In the $\mathbb{Z}_2$ symmetry defects, an inactive link and the inactive sites are doubled. Conversely, when the midpoint of an active link lies within the $\mathbb{Z}_2$ symmetry defect, the active link is doubled, but the two active sites at the ends of this active link remain undoubled. Each triangle on this surface is doubled and becomes a triangular prism due to this deformation (see Figs. 4.7, 4.8). We use this triangular prism as a building block for $\mathbb{Z}_2$ 1-form symmetry defects.

We assign specific weights to $\mathbb{Z}_2$ symmetry defect elements as outlined below. A building block contains the midpoints of two active links. Each building block includes the midpoints of two active links, denoted by $b, c = 0, 1$. Our goal is to define the 1-form $\mathbb{Z}_2$ symmetry defect to flip the sign of a Wilson loop. Therefore, we assign the weight $Z_2(b, c) = \sigma^x_{b,c} = (1 - \delta_{b,c})$ to each building block. We also assign a weight $z = \sqrt{2}$ to each pair of doubled active links, balancing the additional weight (4.3.9) for the doubled active link. The weights for these components are summarized below.

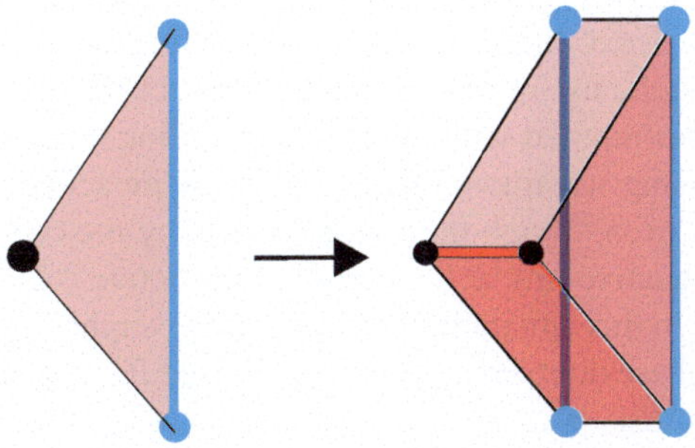

Fig. 4.7 The building block of the 1-form $\mathbb{Z}_2$ center symmetry defect. A black dot represents the midpoint of an active link. A blue dot represents an inactive site and a blue line represents an inactive link. A building block of the surface on which a 1-form $\mathbb{Z}_2$ symmetry defect is supported is a triangle formed by an inactive link and the midpoint of an adjacent active link. A triangle on which a $\mathbb{Z}_2$ symmetry defect is supported is doubled and becomes a triangular prism. Reprinted from [1], Copyright 2021, The Author(s) licensed under CC BY 4.0

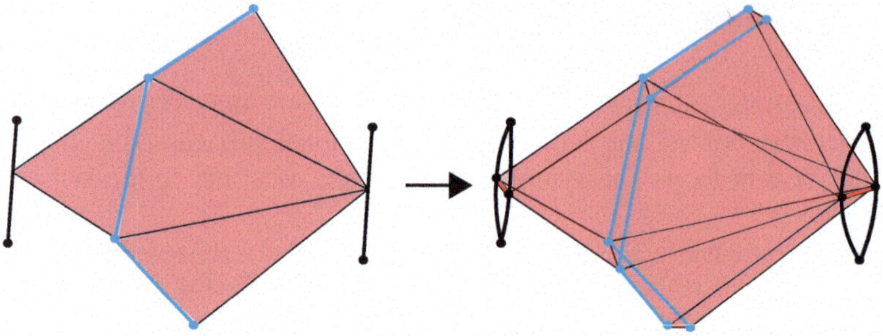

Fig. 4.8 The surface is made by connecting several building blocks of the 1-form $\mathbb{Z}_2$ symmetry defect. A black line represents an active link. A blue dot represents an inactive site and a blue line represents an inactive link. Reprinted from [1], Copyright 2021, The Author(s) licensed under CC BY 4.0

$Z_2(b, c) =$ 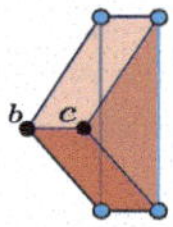

$= \sigma^x_{b,c}$, $z =$ 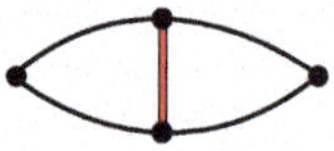 $= \sqrt{2}$. (4.3.15)

(Reprinted from [1], Copyright 2021, The Author(s) licensed under CC BY 4.0.)

Assigning the Boltzmann weight to a plaquette (or a 16-cell) with doubled active links is not entirely clear. The assignment is based on the rules below. Each building block in a $\mathbb{Z}_2$ symmetry defect contains a pair of inactive links that are doubled. In our setup, an inactive link passes through the middle of a cube. This cube is made up of six plaquettes, and we refer to it as an "active cube." Furthermore, the building block for the $\mathbb{Z}_2$ symmetry defect defines the center point of an active link within this active cube. Let's select such a building block and describe the method to assign the Boltzmann weights. In this active cube, there are two plaquettes that share this active link. Next, the Boltzmann weight for one of these two plaquettes is determined through the use of one of the doubled active links, and the weight for the other plaquette is computed with the other active link (refer to Figs. 4.9, 4.10). It could get a bit confusing when two plaquettes inside an active cube share a pair of doubled active links, even though the triangle made by the center of these doubled active links and the inactive link at the center of the active cube is not included in a $\mathbb{Z}_2$ symmetry defect. In this situation, the Boltzmann weights for both plaquettes are derived using one of these doubled active links. These rules determine the Boltzmann weights for all plaquettes.

Presently, we demonstrate that these $\mathbb{Z}_2$ 1-form symmetry defects possess topological properties. We focus on a 16-cell, as with the duality defects. The set of active links in this 16-cell is represented as $M = \{m = 1, 2, 3, 4\}$, while the set of the inactive links is denoted by $\tilde{N} = \{\tilde{n} = \tilde{1}, \tilde{2}, \tilde{3}, \tilde{4}\}$. Let's pick two adjacent active links, $m = 1, 2$, from the set M. Next, we focus on the set of triangles

$$V = \{(m, \tilde{n}) | m = 1, 2,\ \tilde{n} = \tilde{1}, \tilde{2}, \tilde{3}, \tilde{4}\}. \tag{4.3.16}$$

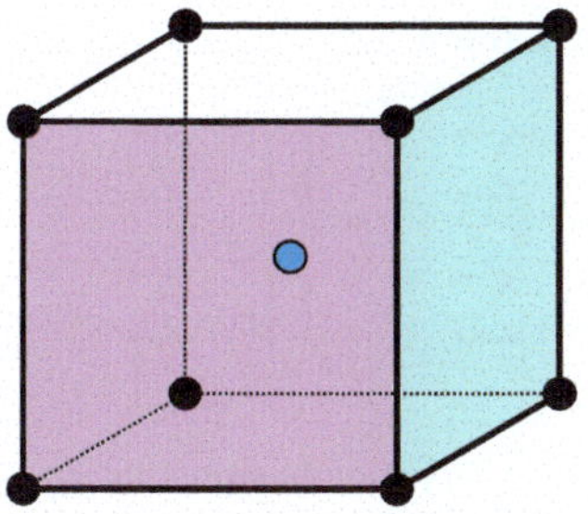

Fig. 4.9 An active cube. A black line represents an active link. A blue dot represents the inactive link crossing the center of the active cube. Reprinted from [1], Copyright 2021, The Author(s) licensed under CC BY 4.0

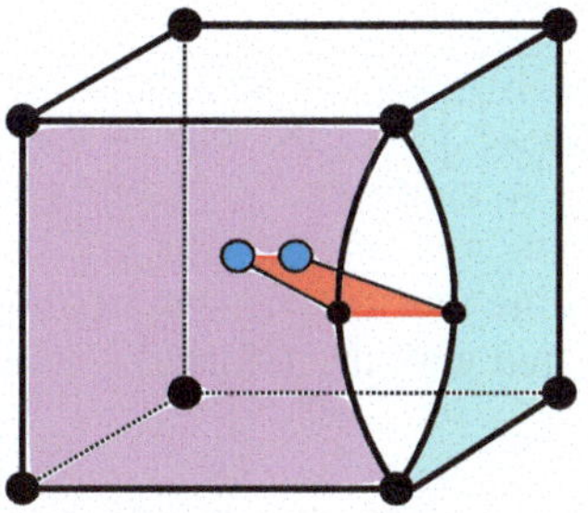

Fig. 4.10 An active cube with a building block of a $\mathbb{Z}_2$ symmetry defect. The inactive link and an active link is doubled by the $\mathbb{Z}_2$ symmetry defect. Two plaquettes that share the active link before doubling contain the doubled active links, one each. Reprinted from [1], Copyright 2021, The Author(s) licensed under CC BY 4.0

These triangles form an octahedron. We examine a $\mathbb{Z}_2$ 1-form symmetry defect made of $H \subset V$ in the octahedron; this configuration is referred to as B. We compare it with the deformed configuration C, where the defect is formed by $\tilde{H} := V \setminus H$ within the octahedron, and the configuration outside the octahedron remains unchanged. In order to write down the commutation relations, we also define $M' = \{m = 1, 2\} \subset M$, $F_b := \{m|(m, \tilde{n}) \in \tilde{H}\}$ and $F_c := \{m|(m, \tilde{n}) \in H\}$. The link variables of the active links 3, 4 are denoted by a_3, a_4, respectively. The link variables of the links $m = 1, 2$ are denoted by b_m and c_m if the link m is doubled. If $m = 1, 2$ is not doubled, its link variable is denoted by b_m in the configuration B and c_m in the configuration C. Then our $\mathbb{Z}_2$ symmetry defects turn out to satisfy the commutation relation

$$\sum_{M' \setminus F_b} W(b_1, b_2, a_3, a_4) \prod_{(m,\tilde{n}) \in H} Z_2(b_m, c_m) = \sum_{M' \setminus F_c} W(c_1, c_2, a_3, a_4) \prod_{(m,\tilde{n}) \in \tilde{H}} Z_2(b_m, c_m). \tag{4.3.17}$$

Here, considering the weight of z, the weights of sites and links on each side are equal and cancel to each other. Therefore, we have not included them in Eq. (4.3.17).

The sums in Eq. (4.3.17) are defined by

$$\sum_{M'\setminus F_b} := \prod_{m\in M'\setminus F_b} \sum_{b_m=0,1}, \tag{4.3.18}$$

$$\sum_{M'\setminus F_c} := \prod_{m\in M'\setminus F_c} \sum_{c_m=0,1}. \tag{4.3.19}$$

The commutation relations (4.3.17) means that our $\mathbb{Z}_2$ symmetry defects are topological.

The commutation relations (4.3.17) also imply our $\mathbb{Z}_2$ symmetry defects are invertible defect. For instance, a $\mathbb{Z}_2$ symmetry defect placed on the octahedron carries the same weight as an empty configuration

$$\sum_{b_1,b_2=0}^{1} W(b_1, b_2, a_3, a_4) \prod_{(m,\tilde{n})\in V} Z_2(b_m, c_m) = W(c_1, c_2, a_3, a_4). \tag{4.3.20}$$

Our study leads us to determine that our $\mathbb{Z}_2$ symmetry defects truly correspond to the symmetry defects associated with the 1-form $\mathbb{Z}_2$ center symmetry.

4.3.3 *Defect Junctions*

We can think about a setup of defects where various types of defects encounter each other, creating junctions. This part of the discussion focuses on such junctions and their weight.

In our model, a junction happens when a $\mathbb{Z}_2$ 1-form symmetry defect ends on a duality defect, as shown in Fig. 4.11. These junctions are found on one-dimensional lines.

As explained in Sect. 4.3.1, duality defects are established by doubling tetrahedrons and their associated links. A junction forms when a doubled link of the duality defect intersects with the $\mathbb{Z}_2$ symmetry defect. There are two types of junctions,

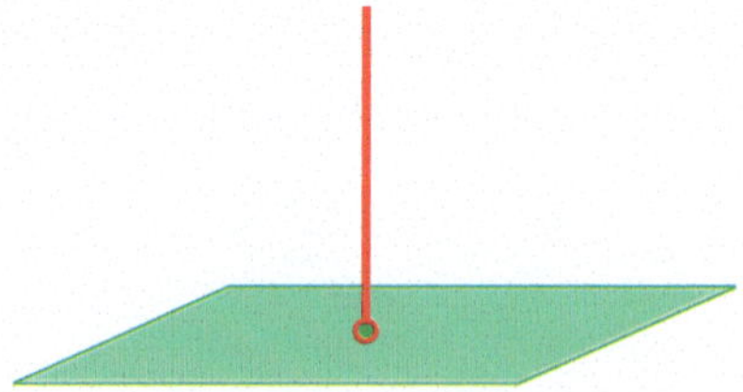

Fig. 4.11 A schematic illustration of junctions. The red line represents a $\mathbb{Z}_2$ symmetry defect, and the green surface represents a duality defect. The intersection is actually 1-dimensional. Reprinted from [1], Copyright 2021, The Author(s) licensed under CC BY 4.0

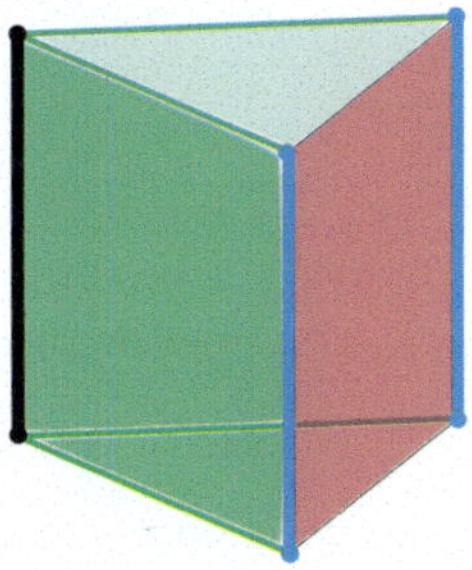

Fig. 4.12 A schematic illustration of a junction sharing an inactive link. Its weight is denoted as $J(a)$. Reprinted from [1], Copyright 2021, The Author(s) licensed under CC BY 4.0

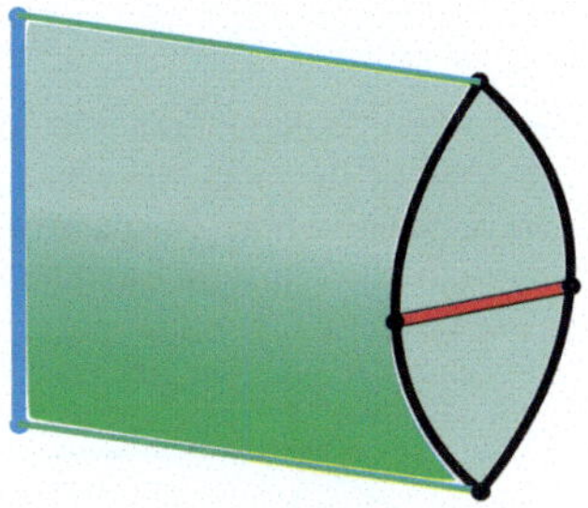

Fig. 4.13 A schematic illustration of a junction sharing an active link. Its weight is denoted as $\tilde{J}(b, c)$. Reprinted from [1], Copyright 2021, The Author(s) licensed under CC BY 4.0

based on whether the shared link is an active lattice or an inactive lattice. Each type of junction is depicted as in Figs. 4.12 and 4.13. The weight assigned to a junction that shares an inactive link is represented by $J(a)$, using the link variable a for the active link that matches the shared inactive link. In the same way, if the shared link is an active link, the weight of the resulting junction is indicated by $\tilde{J}(b, c)$, where b, c are the link variables of the shared links that are doubled by the $\mathbb{Z}_2$ symmetry defect. The total weight becomes non-zero only when $b = 1 - c$, a result of the $\mathbb{Z}_2$ symmetry defect. Therefore, we can apply a function $\tilde{J}(a)$, depending only on a, to represent the weight of this junction as $\tilde{J}(b, c) = \tilde{J}(b)\sigma^x_{b,c}$.

The weights at these junctions are set to ensure that the following junction commutation relations are met. To describe junction commutation relations in detail, we use a diagram as illustrated in Fig. 4.14. In this illustration, every corner symbolizes a tetrahedron. Two tetrahedrons connected with a black line share an active link, and two tetrahedrons connected with a blue line share an inactive link.

First, we apply a junction commutation relation to the J junction weights and the $\tilde{J}$ junction weights. We examine two configurations of defects as shown in Fig. 4.15, ensuring their weights are identical. This commutation relation demands that the $\mathbb{Z}_2$ symmetry defect is capable of continuous transformation along with the duality defect, as illustrated in Fig. 4.16. This commutation relation can be expressed as

$$\begin{aligned}
&W(1-a_1, a_2, a_3, a_4)D(1-a_1, \tilde{a}_{\bar{2}})D(1-a_1, \tilde{a}_{\bar{3}})D(a_2, \tilde{a}_{\bar{1}})D(a_2, \tilde{a}_{\bar{2}})D(a_2, \tilde{a}_{\bar{3}})J(\tilde{a}_{\bar{1}})\tilde{J}(1-a_1) \\
&= W(a_1, 1-a_2, a_3, a_4)D(a_1, \tilde{a}_{\bar{2}})D(a_1, \tilde{a}_{\bar{3}})D(1-a_2, \tilde{a}_{\bar{1}})D(1-a_2, \tilde{a}_{\bar{2}})D(1-a_2, \tilde{a}_{\bar{3}})\tilde{J}(1-a_2).
\end{aligned} \tag{4.3.21}$$

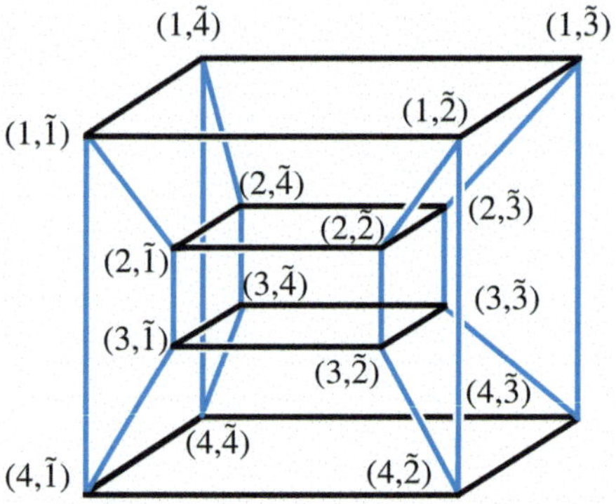

Fig. 4.14 A configuration of tetrahedrons. Each vertex represents a tetrahedron. Two tetrahedrons connected by the black line share an active link, and two tetrahedrons connected by the blue line share an inactive link. The pairs of numbers assigned to the vertices specify the labels of the active and inactive links that each tetrahedron contains. Reprinted from [1], Copyright 2021, The Author(s) licensed under CC BY 4.0

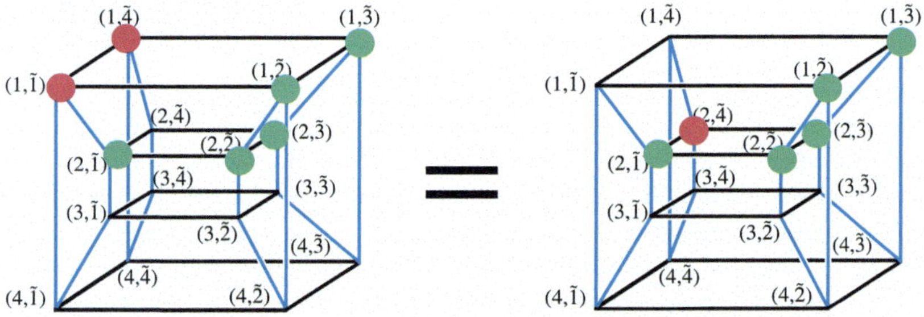

Fig. 4.15 The configuration of Eq. (4.3.21). The green dots represent the tetrahedrons on which the duality defect is located. The red dots represent the tetrahedrons on which the $\mathbb{Z}_2$ symmetry defect is located. Here, placing a $\mathbb{Z}_2$ symmetry defect on a tetrahedron means placing a $\mathbb{Z}_2$ symmetry defect on a triangle connecting the center of the active link with the inactive link in the tetrahedron. Reprinted from [1], Copyright 2021, The Author(s) licensed under CC BY 4.0

In this case, the constant weights are not included as they are the same for both parts. The summation arising from the $\mathbb{Z}_2$ symmetry defect have already been removed.

Next, we determine the $\tilde{J}(a, b)$ junction weight using junction commutation relations. We examine two defect configurations depicted in Fig. 4.17, ensuring their weights are identical. The junction commutation relation is also a requirement that the $\mathbb{Z}_2$ symmetry defect can be continuously deformed along the duality defect. This equation reads

$$\begin{aligned} &W(1-a_1, a_2, a_3, a_4)D(1-a_1, \tilde{a}_{\tilde{2}})D(1-a_1, \tilde{a}_{\tilde{3}})D(a_2, \tilde{a}_{\tilde{2}})D(a_2, \tilde{a}_{\tilde{3}})\tilde{J}(1-a_1) \\ &= W(a_1, 1-a_2, a_3, a_4)D(a_1, \tilde{a}_{\tilde{2}})D(a_1, \tilde{a}_{\tilde{3}})D(1-a_2, \tilde{a}_{\tilde{2}})D(1-a_2, \tilde{a}_{\tilde{3}})\tilde{J}(a_1)\tilde{J}(1-a_2). \end{aligned} \tag{4.3.22}$$

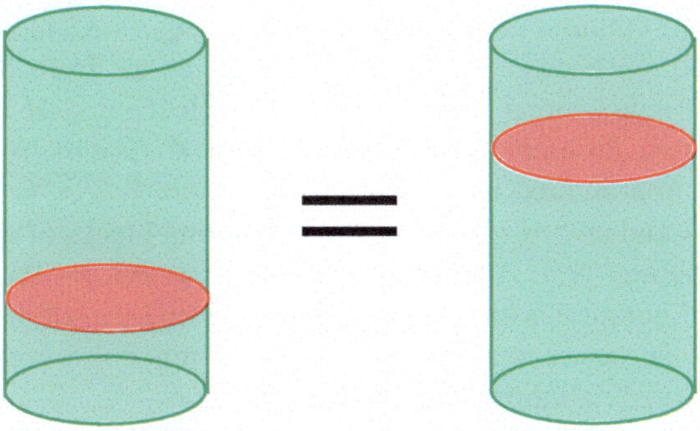

Fig. 4.16 A schematic illustration of junction commutation relations. The green surface represents a duality defect and the red surface represents a $\mathbb{Z}_2$ symmetry defect. The junction commutation relation requires that the $\mathbb{Z}_2$ defect can be deformed along with the duality defect. Junctions are located on the intersection of the two kinds of defects.

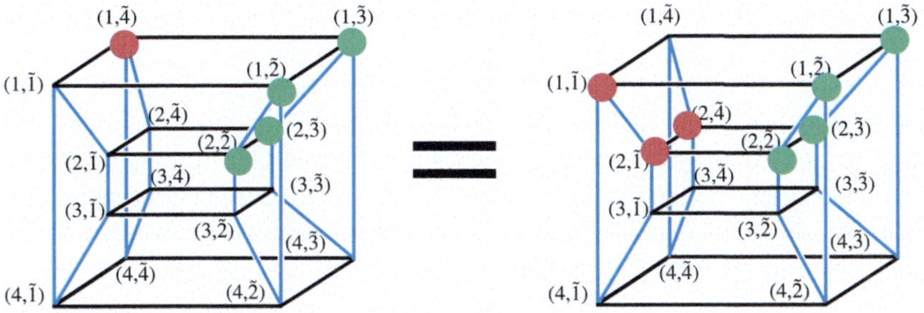

Fig. 4.17 The configuration of Eq. (4.3.22).

In this case, the constant weights are not included as they are the same for both parts. The summation arising from the $\mathbb{Z}_2$ symmetry defect have already been removed.

We can resolve the junction commutation relations from Eqs. (4.3.21) and (4.3.22), and discover a unique solution for $J(a)$, $\tilde{J}(b, c)$. It is written as

$$J(a) = \; ^{a}\,[\text{diagram}] = (-1)^a, \tag{4.3.23}$$

$$\tilde{J}(b, c) = [\text{diagram}] = \sigma^x_{b,c}. \tag{4.3.24}$$

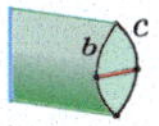

General junction commutation relations include Eqs. (4.3.21) and (4.3.22). We describe these commutation relations using the notations from Sect. 4.3.1. Let's assume that duality defects are placed on $I \subset U$. We select $i, j \in M = \{1, 2, 3, 4\} (i \neq j)$ and examine two different configurations with $\mathbb{Z}_2$ symmetry defects. These configurations are represented by $F = i, j \times \tilde{N},\ \tilde{N} := \tilde{1}, \tilde{2}, \tilde{3}, \tilde{4}$. We define $I_F = \{(m, \tilde{n}) | (m, \tilde{n}) \in I \text{ and } m \in \{i, j\}\}$ as the set of building blocks of the duality defect with active links being i or j. Additionally, we set $\bar{F} = F \setminus I_F$. We require that the weight of the configuration with a $\mathbb{Z}_2$ defect placed in $H_F \subset \bar{F}$ must be the same as that in $\bar{F} \setminus H_F$. These are non-trivial relations for junction weights. The weights (4.3.23), (4.3.24) satisfy all these commutation relations.

Let's describe why our junctions remain unchanged under gauge transformations. The expression (4.3.24) for weight is unaffected by gauge transformations. This happens because at an active site within this building block, it changes as $\tilde{J}(b, c) \to \tilde{J}(1 - b, 1 - c) = \tilde{J}(b, c)$ due to the gauge transformation. On the other hand, $J(a)$ in Eq. (4.3.23) is not invariant under gauge transformation by itself, as it changes to $J(a) \to J(1 - a) = -J(a)$ due to the gauge transformation at an active site in this building block of junctions. However a whole junction remains gauge invariant as described. Let's concentrate on an active site S on a J junction. Since this junction is a line with no boundary, there are precisely two building blocks of junctions which include S. Let a, b be the link variables for the two active links in these two building blocks. By the gauge transformation at S, the weights of these two building blocks change as $J(a)J(b) \to J(1 - a)J(1 - b) = J(a)J(b)$. The other part of the junction remains unchanged by this gauge transformation. Thus, we can conclude that our junctions are gauge invariant.

4.3.4 Crossing Relations and Expectation Values

In this subsection, we explore different crossing relationships using the defects and junctions explained so far. For example, we derive the crossing relations between two duality defect configurations whose topologies are different. We further deduce certain crossing relations where both duality defects and $\mathbb{Z}_2$ symmetry defects emerge.

A demonstration of such relations is that $\mathbb{Z}_2$ symmetry defects having a boundary on the duality defect can be eliminated if the boundary is homologically trivial on the duality defect. It is schematically illustrated in Fig. 4.18. One concrete configuration in the lattice is shown in Fig. 4.19. It's possible to confirm the equation shown in Fig. 4.19.

$$\sum_{a_1, b_3 = 0,1} J(\tilde{a}_{\tilde{1}})J(\tilde{a}_{\tilde{2}})J(\tilde{a}_{\tilde{3}})J(\tilde{a}_{\tilde{4}})W(a_1, a_2, b_3, a_4)Z_2(b_3, c_3)^4 D(a_1, \tilde{a}_{\tilde{1}})D(a_1, \tilde{a}_{\tilde{2}})D(a_1, \tilde{a}_{\tilde{3}})D(a_1, \tilde{a}_{\tilde{4}})$$
$$= \sum_{a_1 = 0,1} W(a_1, a_2, c_3, a_4)D(a_1, \tilde{a}_{\tilde{1}})D(a_1, \tilde{a}_{\tilde{2}})D(a_1, \tilde{a}_{\tilde{3}})D(a_1, \tilde{a}_{\tilde{4}}). \tag{4.3.25}$$

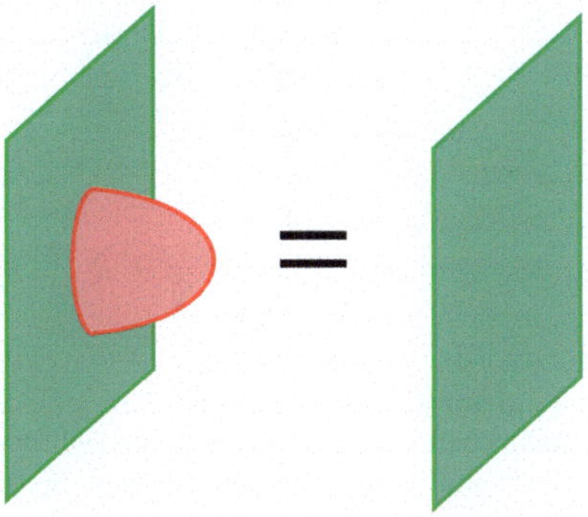

Fig. 4.18 A schematic illustration of the crossing relations that a $\mathbb{Z}_2$ symmetry defects with a boundary on the duality defect can be removed if the boundary is homologically trivial on the duality defect. Reprinted from [1], Copyright 2021, The Author(s) licensed under CC BY 4.0

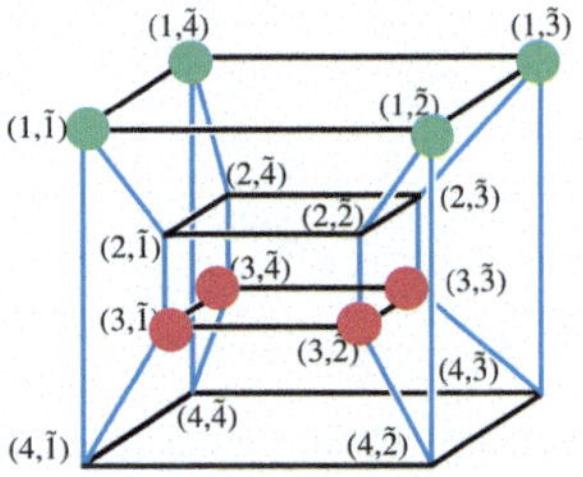

Fig. 4.19 The configuration of the left-hand side of Eq. (4.3.25). Reprinted from [1], Copyright 2021, The Author(s) licensed under CC BY 4.0

Here, we omit the constant weights like those of links, as these weights are identical on both sides.

Another crucial intersection relation is the formula for the duality defect on a 3-dimensional solid torus. We think about splitting a 16-cell into a pair of solid tori. These solid tori are referred to as V_1 and V_2. A solid torus is not simple connected, and thus the topology of the duality defects changes by the deformation as explained in Sect. 4.3.1. Therefore, it's not necessary for the defect commutation relations to be satisfied. Instead, we find the crossing relations of the solid torus that include $\mathbb{Z}_2$ defects (refer to Fig. 4.20); there is a connection between the duality defect on V_1, the duality defect on V_2, and the duality defect on V_2 along with the $\mathbb{Z}_2$ symmetry defect. Define I_S as the collection of tetrahedrons in V_2. First, we select one $\tilde{k}$ from $N = \{\tilde{1}, \tilde{2}, \tilde{3}, \tilde{4}\}$, and define $I_S^{\tilde{k}} := \{(m, \tilde{k}) | (m, \tilde{k}) \in I_S\}$ as the set of tetrahedrons in V_2 that contain $\tilde{k}$. Think about a duality defect on V_2 and a $\mathbb{Z}_2$ symmetry defect on each tetrahedron in $M \times \tilde{k} \setminus I_S^{\tilde{k}},\ M = 1, 2, 3, 4$. Here, placing the $\mathbb{Z}_2$ symmetry defect on the tetrahedron refers to situating the $\mathbb{Z}_2$ symmetry defect within a triangle in a way that it joins the middle of the active link to the inactive link inside the tetrahedron. Notice that the weight of the configuration is independent of the choice $\tilde{k}$ because of the defect commutation relations described in Sect. 4.3.3. For example, if the duality defect is placed on V_1 as shown in the left-hand side of Fig. 4.21, the configuration

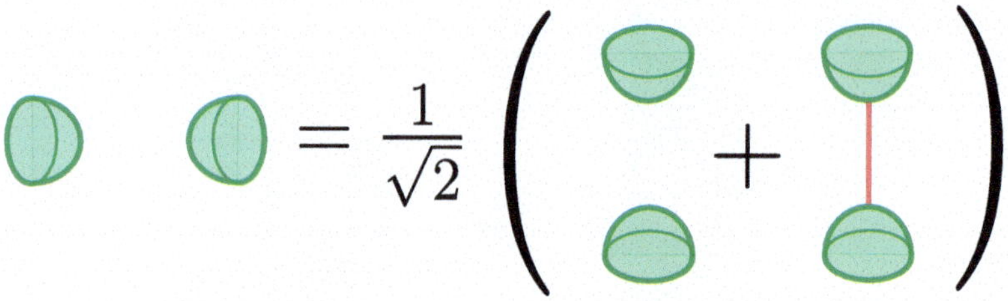

Fig. 4.20 A schematic illustration of solid torus equations. The left-hand side represents a duality defect configuration on one solid torus V_1. The first term of the right-hand side represents a duality defect configuration on one solid torus V_2 and the second term represents a configuration of the duality defect on a solid torus V_2 with a $\mathbb{Z}_2$ symmetry defect on D^2 whose boundary is a non-trivial cycle on the duality defect. This is an example in which the crossing relation is not closed only within duality defects. Reprinted from [1], Copyright 2021, The Author(s) licensed under CC BY 4.0

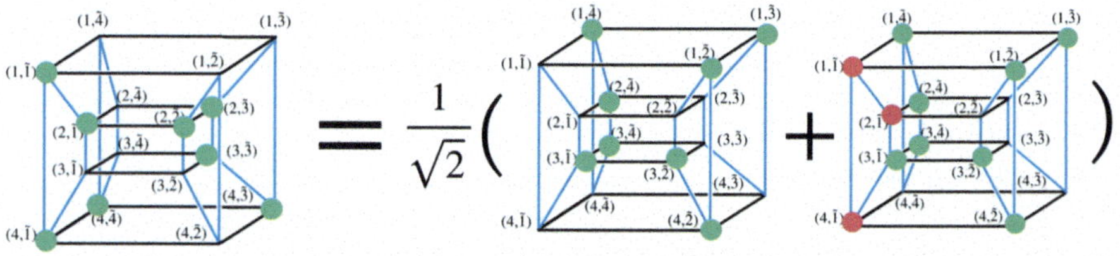

Fig. 4.21 The configuration of Eq. (4.3.26). The left-hand side is the duality defect configuration on V_1. The first term on the right-hand side is the duality defect configuration on V_2, and the second term on the right-hand side is the duality defect configuration on V_2 with a $\mathbb{Z}_2$ symmetry defect. Reprinted from [1], Copyright 2021, The Author(s) licensed under CC BY 4.0

of the $\mathbb{Z}_2$ symmetry defect is shown in the second term on the right-hand side of Fig. 4.21.

Next, the crossing relation depicted in Fig. 4.21 holds true. The specific equation for this is as follows.

$$\begin{aligned}
& s^8 l^8 W(a_1, a_2, a_3, a_4) D(a_1, \tilde{a}_{\bar{1}}) D(a_2, \tilde{a}_{\bar{1}}) D(a_2, \tilde{a}_{\bar{2}}) D(a_2, \tilde{a}_{\bar{3}}) D(a_3, \tilde{a}_{\bar{3}}) D(a_4, \tilde{a}_{\bar{1}}) D(a_4, \tilde{a}_{\bar{3}}) D(a_4, \tilde{a}_{\bar{4}}) \\
& = \frac{1}{\sqrt{2}} \Big(s^8 l^8 W(\tilde{a}_{\bar{1}}, \tilde{a}_{\bar{2}}, \tilde{a}_{\bar{3}}, \tilde{a}_{\bar{4}}) D(a_1, \tilde{a}_{\bar{2}}) D(a_1, \tilde{a}_{\bar{3}}) D(a_1, \tilde{a}_{\bar{4}}) D(a_2, \tilde{a}_{\bar{4}}) D(a_3, \tilde{a}_{\bar{1}}) D(a_3, \tilde{a}_{\bar{2}}) D(a_3, \tilde{a}_{\bar{4}}) D(a_4, \tilde{a}_{\bar{2}}) \\
& + \sum_{\tilde{b}_{\bar{1}}=0,1} s^8 l^9 z W(\tilde{b}_{\bar{1}}, \tilde{a}_{\bar{2}}, \tilde{a}_{\bar{3}}, \tilde{a}_{\bar{4}}) D(a_1, \tilde{a}_{\bar{2}}) D(a_1, \tilde{a}_{\bar{3}}) D(a_1, \tilde{a}_{\bar{4}}) D(a_2, \tilde{a}_{\bar{4}}) D(a_3, \tilde{b}_{\bar{1}}) D(a_3, \tilde{a}_{\bar{2}}) \\
& \quad \times D(a_3, \tilde{a}_{\bar{4}}) D(a_4, \tilde{a}_{\bar{2}}) Z_2(\tilde{a}_{\bar{1}}, \tilde{b}_{\bar{1}})^3 J(a_1) J(a_2) J(a_3) \tilde{J}(\tilde{a}_{\bar{1}}, \tilde{b}_{\bar{1}}) \Big).
\end{aligned} \tag{4.3.26}$$

Using the solid torus equations (4.3.26), we are able to calculate several expectation values. The first example is the expectation value of a duality defect on $S^2 \times S^1$, as illustrated in Fig. 4.22. We can cut out one solid torus from $S^2 \times S^1$. We apply the solid torus equation to this solid torus. Using the connections in Figs. 4.6 and 4.18, we understand that the expectation value of the duality defect on $S^2 \times S^1$ is one. The second example is an expectation value of a $\mathbb{Z}_2$ symmetry defect on a two-

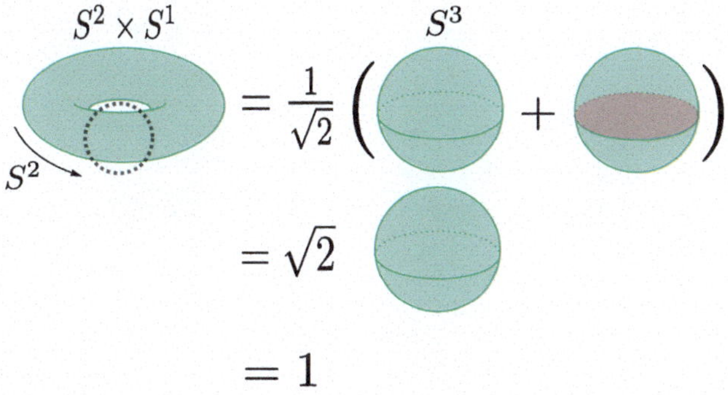

Fig. 4.22 A calculation of an expectation value of a duality defect on $S^2 \times S^1$. In the left-hand side, we focus on the part surrounded by the dotted circle. The duality defect in this part is $D^2 \times S^1$ shaped, and therefore we apply the solid torus equation here. In the right-hand side in the first line, the first term is duality defect on S^3. The second term is a duality defect on S^3 and a $\mathbb{Z}_2$ symmetry defect on D^2 whose boundary is on the duality defect. By using a crossing relation in Fig. 4.18, we find the second term equal to the first term. Therefore, the expectation value of a duality defect on $S^2 \times S^1$ is equal to the $\sqrt{2}$ times S^3 expectation value. By using the relation in Fig. 4.6, finally, we find the expectation value of a duality defect on $S^2 \times S^1$ is one. Reprinted from [1], Copyright 2021, The Author(s) licensed under CC BY 4.0

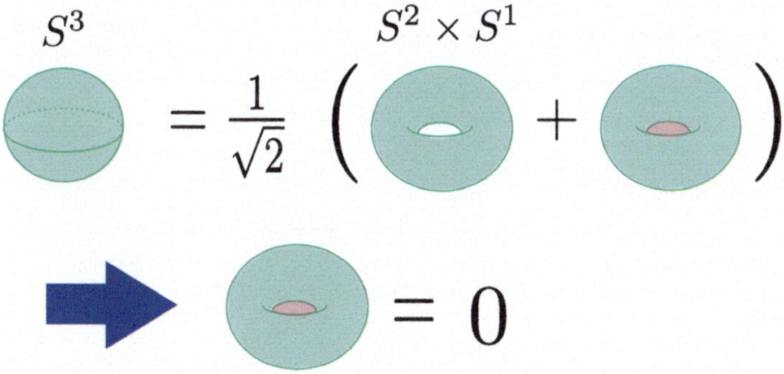

Fig. 4.23 A calculation of the expectation value of a duality defect on $S^2 \times S^1$ with a $\mathbb{Z}_2$ symmetry defect on D^2 whose boundary is a non-trivial cycle on the duality defect. In the first equality, we use the solid torus equation in Fig. 4.20. By using the relations in Figs. 4.6 and 4.22, we find that the expectation value is zero. Reprinted from [1], Copyright 2021, The Author(s) licensed under CC BY 4.0

dimensional disk, which has a boundary that is a non-trivial S^1 cycle on the duality defect on $S^2 \times S^1$, as depicted in Fig. 4.23. We apply the solid torus equation to the duality defect configuration on S^3. Then, we find that the expectation value is zero.

Additionally, We also find the crossing relation for duality defects placed on two disconnected 3-dimensional disks as depicted in Fig. 4.24. This relation is realized in a 16-cell as illustrated in Fig. 4.25. This connection is described as

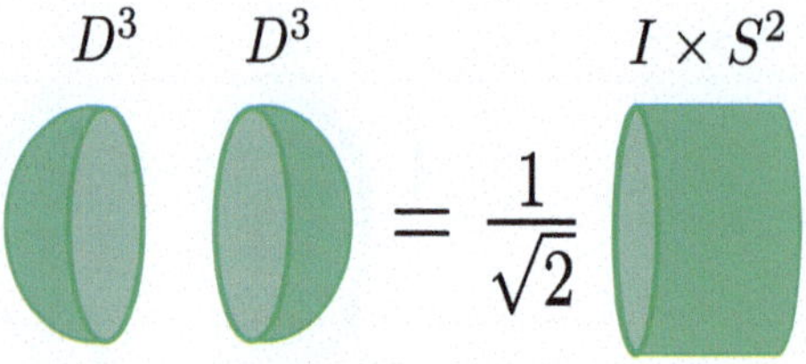

Fig. 4.24 A schematic illustration of the crossing relation including two disconnected disks. Reprinted from [1], Copyright 2021, The Author(s) licensed under CC BY 4.0

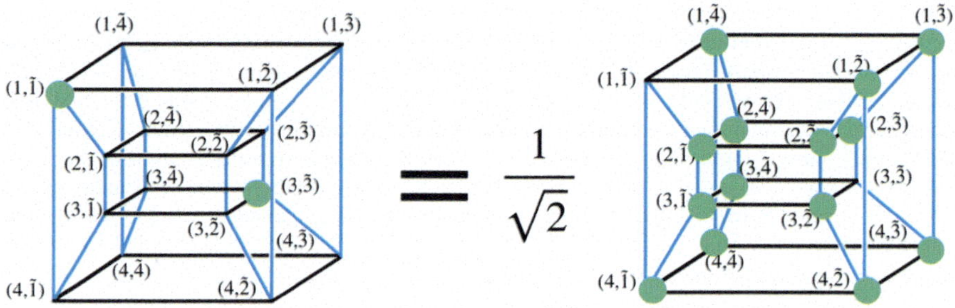

Fig. 4.25 The configuration of Eq. (4.3.27). Reprinted from [1], Copyright 2021, The Author(s) licensed under CC BY 4.0

$$
\begin{aligned}
& s^8 l^6 W(a_1, a_2, a_3, a_4) D(a_1, \tilde{a}_{\bar{1}}) D(a_3, \tilde{a}_{\bar{3}}) \\
&= \frac{1}{\sqrt{2}} \sum_{\tilde{a}_2, \tilde{a}_4 = 0,1} s^8 l^8 W(\tilde{a}_{\bar{1}}, \tilde{a}_{\bar{2}}, \tilde{a}_{\bar{3}}, \tilde{a}_{\bar{4}},) D(a_1, \tilde{a}_{\bar{2}}) D(a_1, \tilde{a}_{\bar{3}}) D(a_1, \tilde{a}_{\bar{4}}) D(a_2, \tilde{a}_{\bar{1}}) D(a_2, \tilde{a}_{\bar{2}}) D(a_2, \tilde{a}_{\bar{3}}) \\
&\quad \times D(a_2, \tilde{a}_{\bar{4}}) D(a_3, \tilde{a}_{\bar{1}}) D(a_3, \tilde{a}_{\bar{2}}) D(a_3, \tilde{a}_{\bar{4}}) D(a_4, \tilde{a}_{\bar{1}}) D(a_4, \tilde{a}_{\bar{2}}) D(a_4, \tilde{a}_{\bar{3}}) D(a_4, \tilde{a}_{\bar{4}}).
\end{aligned}
\tag{4.3.27}
$$

This relation allows us to compute the expectation values of the duality defects. An example is a duality defect on S^3. We think about two disconnected duality defects on S^3. We can establish a connection between this configuration and a S^3 duality defect using Eq. (4.3.27). Consequently, the expectation value of the duality defect on S^3 is $1/\sqrt{2}$. This result is consistent with the result in Eq. (4.3.14). The second example is a duality defect on $S^1 \times S^2$ as shown in Fig. 4.26. Using the crossing relation in Fig. 4.24 to $S^1 \times S^2$, we find that the $S^1 \times S^2$ expectation value is one.

In a more general sense, one can compute the expectation values of the duality defect on the connected sum by utilizing the crossing relationship depicted in Fig. 4.24. Let $\langle X \rangle$ be the expectation values of the duality defects on a sub-manifold X. Then the expectation value of the connected sum is provided by

$$\langle X \# Y \rangle = \sqrt{2} \langle X \rangle \langle Y \rangle . \tag{4.3.28}$$

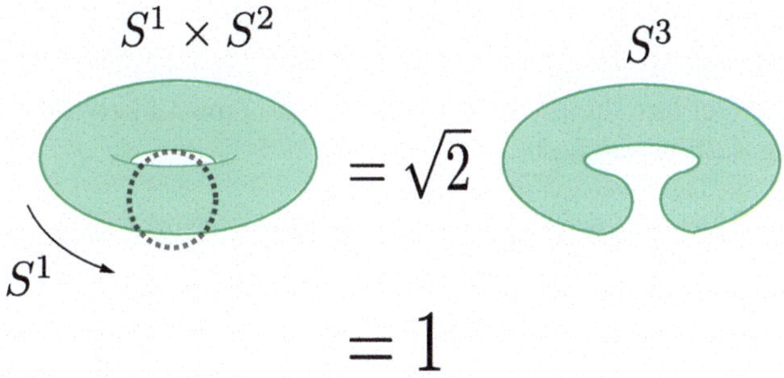

Fig. 4.26 A calculation of an expectation value of a duality defect on $S^1 \times S^2$. We use the commutation relation Fig. 4.24 to $S^1 \times S^2$. Then, the $S^1 \times S^2$ expectation value is equal to $\sqrt{2}$ times the S^3 expectation value. We see that the $S^1 \times S^2$ expectation value is one by the relation Fig. 4.6. Reprinted from [1], Copyright 2021, The Author(s) licensed under CC BY 4.0

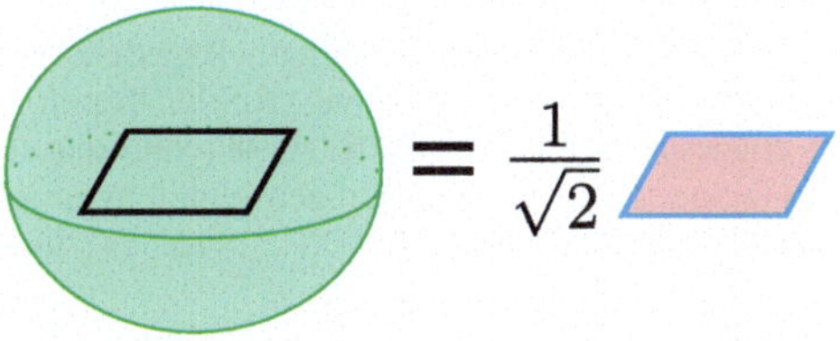

Fig. 4.27 A schematic illustration of the action of a duality defect to a Wilson loop. Reprinted from [1], Copyright 2021, The Author(s) licensed under CC BY 4.0

For example, $\langle (S^1 \times S^2)\#(S^1 \times S^2) \rangle = \sqrt{2}$.

We describe the action of a duality defect on a Wilson loop. When we place the duality defect around a Wilson loop on a plaquette, the relation in Fig. 4.27 is derived; a 't Hooft loop appears on the plaquette where the Wilson loop was placed. In this theory, a 't Hooft loop itself cannot be defined locally within the loop. It must be accompanied with a $\mathbb{Z}_2$ symmetry defect on a surface whose boundary is the loop [6]. The relation in Fig. 4.27 is expressed as

$$\sum_{\tilde{a}_{\tilde{1}},\tilde{a}_{\tilde{2}},\tilde{a}_{\tilde{3}},\tilde{a}_{\tilde{4}}=0,1} W(a_1, a_2, a_3, a_4) s^8 l^8 (-1)^{\tilde{a}_{\tilde{1}}+\tilde{a}_{\tilde{2}}+\tilde{a}_{\tilde{3}}+\tilde{a}_{\tilde{4}}} \prod_{(m,\tilde{n})\in U} D(a_m, \tilde{a}_{\tilde{n}}) = \frac{1}{\sqrt{2}} W(1-a_1, a_2, a_3, a_4) s^4 l^4. \tag{4.3.29}$$

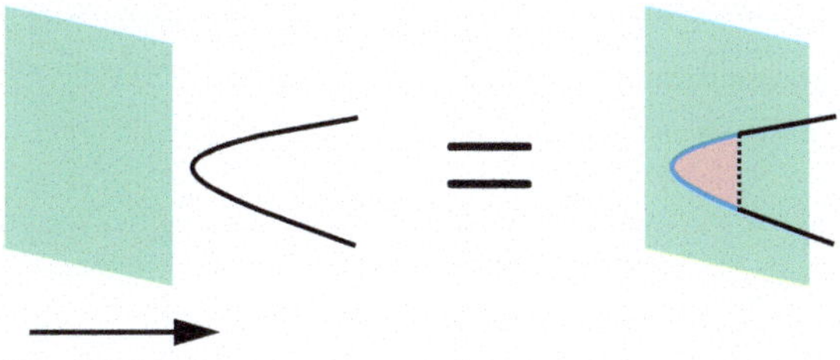

Fig. 4.28 A schematic illustration of the general action of a duality defect to a Wilson loop. Reprinted from [1], Copyright 2021, The Author(s) licensed under CC BY 4.0

In this equation, U is defined as $U = \{1, 2, 3, 4\} \times \{\tilde{1}, \tilde{2}, \tilde{3}, \tilde{4}\}$. The right-hand side $1/\sqrt{2}$ represents the expectation value of the duality defect on S^3. The sum of the $\mathbb{Z}_2$ symmetry defect has already been evaluated. It is most likely that more general relations in Fig. 4.28 is satisfied.

References

1. Koide, M., Nagoya, Y., Yamaguchi, S.: Non-invertible topological defects in 4- dimensional Z2 pure lattice gauge theory. PTEP **2022**(1), 013B03 (2022). arXiv: 2109.05992 [hep-th]
2. Aasen, D., Mong, R.S.K., Fendley, P.: Topological defects on the lattice I: the ising model. J. Phys. A **49**(35), 354001 (2016). https://doi.org/10.1088/1751-8113/49/35/354001. arXiv: 1601.07185 [cond-mat.stat-mech]
3. Coxeter, H.S.M.: Regular Polytopes. Dover Books on Mathematics. Dover Publications (2012). ISBN: 9780486141589
4. 16-cell. in Wikipedia. https://en.wikipedia.org/wiki/16-cell
5. Aasen, D., Fendley, P., Mong, R.S.K.: Topological defects on the lattice: dualities and degeneracies (2020). arXiv: 2008.08598 [cond-mat.stat-mech]
6. Gaiotto, D., et al.: Generalized global symmetries. JHEP **02**, 172 (2015). https://doi.org/10.1007/JHEP02(2015)172. arXiv: 1412.5148 [hep-th]

Chapter 5
Application to G-Functions

In this section, we restrict the renormalization group flow of the theory with boundary using the non-invertible topological defects constructed in the previous section. This restriction is given by the g-theorem.

The hemisphere partition function with conformal boundary conditions is monotonically decreasing in two and three dimensions and increasing in four dimensions, depending on the boundary renormalization group flow. Therefore, once the size of the hemisphere partition function is known, we can place a restriction on the renormalization group flow. In this thesis, we refer to the hemisphere partition function as the g-function.

We compare in this section the g-functions defined by different boundary conditions using non-invertible topological defects. To begin, we first check how the boundary conditions of the theory are defined in our setup. We then define two types of Dirichlet boundary conditions and a Neumann boundary condition. In order to compare g-functions with different boundary conditions, we consider partition functions with defects ending on the boundary and topologically deform the boundary of the defect. Therefore, we define a junction of defects and boundaries that can be topologically deformed. Finally, we compare the g-functions by considering a partition function with a defect that has an edge at the boundary and deforming the defect in two different ways that produce different g-functions. The deviation then arises from the non-invertible nature of the defect. This chapter is based on the following work [1].

M. Koide, *Non-invertible Symmetry in 4-Dimensional Z2 Lattice Gauge Theory*,
Springer Theses, https://doi.org/10.1007/978-981-96-2272-6_5

5.1 Four-Dimensional $\mathbb{Z}_2$ Lattice Gauge Theory with Boundary and the Duality Defects

5.1.1 Boundary Conditions

This subsection introduces three boundary conditions. We define two Dirichlet boundary conditions and one Neumann boundary condition on our setup. We introduce M as a bounded spacetime.

$$M := \{(x_1, x_2, x_3, x_4)|x_1 \geq 0\}. \tag{5.1.1}$$

In this case, the boundary of the spacetime is located at $x = 0$. The boundary, a 3-dimensional lattice at $x = 0$, is part of the lattice Λ introduced in Sect. 4.2. In Sect. 4.2, we restricted the lattice on the Λ to be the active lattice, but in this section we do not fix it to define the boundary conditions.

First, we consider the case of an inactive lattice on Λ. At this time, there are no active links on the boundary, so no restrictions are imposed on the link variables. Therefore, this boundary condition is the free boundary condition, which we define as the Neumann boundary condition and denote by N.

We can assign arbitrary weights to sites, links, and plaquettes on the boundary. Here, for simplicity we fix these weights to 1. Then we can define the partition function of the 4-dimensional $\mathbb{Z}_2$ lattice gauge theory with boundary condition N as follows.

$$Z_{\mathrm{N}} = \sum_{\{a\}} \left(\prod_{\substack{\text{active} \\ \text{sites}}} s \right) \left(\prod_{\substack{\text{active} \\ \text{links}}} l \right) \prod_{i \in C} W(a_{j_1(i)}, a_{j_2(i)}, a_{j_3(i)}, a_{j_4(i)}). \tag{5.1.2}$$

Next we consider the boundary conditions for the case of an active lattice on Λ. In this case we can define two different Dirichlet boundary conditions. The first one is a boundary condition that fixes the link variable a on the boundary to zero. We denote this Dirichlet boundary condition by D. We define the weights of the site, link, and plaquette on the boundary to be s_{D}, l_{D}, and W_{D}, respectively. These values are determined from the conditions for topological deformation of the edge of the duality defect on the boundary in next subsection. The partition function Z_{D} with boundary condition D is as follows.

$$Z_{\mathrm{D}} = \sum_{\{a\}} \left(\prod_{\substack{\text{bulk active} \\ \text{sites}}} s \right) \left(\prod_{\substack{\text{bulk active} \\ \text{links}}} l \right) \left(\prod_{\substack{\text{boundary} \\ \text{sites}}} s_{\mathrm{D}} \right) \left(\prod_{\substack{\text{boundary} \\ \text{links}}} l_{\mathrm{D}} \right) \left(\prod_{\substack{\text{boundary} \\ \text{plaquettes}}} W_{\mathrm{D}} \right) \prod_{i \in C} W(a_{j_1(i)}, a_{j_2(i)}, a_{j_3(i)}, a_{j_4(i)}), \tag{5.1.3}$$

where C is the set of the active plaquettes containing at least one bulk active link, and $j_1(i)$, $j_2(i)$, $j_3(i)$, $j_4(i)$ are active links in active plaquette i. The summation for

$\{a\}$ is taken for all link variable configurations such that the boundary conditions are satisfied.

The second Dirichlet boundary condition is a boundary condition that fixes all active plaquettes on the boundary at zero. That is, any boundary plaquette satisfies $a_1 + a_2 + a_3 + a_4 \equiv 0 \pmod 2$ for the link variables a_i $(i = 1, 2, 3, 4)$ of the active links that comprise the plaquette. We denote this boundary condition by $\widetilde{\mathrm{D}}$. Let $s_{\widetilde{\mathrm{D}}}$ and $l_{\widetilde{\mathrm{D}}}$ be the weights of the active sites and links on the boundary in this boundary condition, respectively. The Boltzmann weight on the boundary is defined as $W_{\widetilde{\mathrm{D}}}\delta^{\mathrm{mod}\ 2}_{a_1+a_2+a_3+a_4,0}$, since the sum of the active link variables must satisfy $a_1 + a_2 + a_3 + a_4 \equiv 0 \pmod 2)=0$ from the boundary condition. Here, $\delta^{\mathrm{mod}\ 2}_{a,0}$ is defined as

$$\delta^{\mathrm{mod}\ 2}_{a,0} := \begin{cases} 0 & (a : \mathrm{odd}) \\ 1 & (a : \mathrm{even}) \end{cases}. \tag{5.1.4}$$

The partition function with boundary condition $\widetilde{\mathrm{D}}$ is given by

$$Z_{\widetilde{\mathrm{D}}} = \sum_{\{a\}} \left(\prod_{\substack{\text{bulk active} \\ \text{sites}}} s \right) \left(\prod_{\substack{\text{bulk active} \\ \text{links}}} l \right) \left(\prod_{\substack{\text{boundary} \\ \text{sites}}} s_{\widetilde{\mathrm{D}}} \right) \left(\prod_{\substack{\text{boundary} \\ \text{links}}} l_{\widetilde{\mathrm{D}}} \right)$$
$$\times \left(\prod_{k \in C_B} W_{\widetilde{\mathrm{D}}} \delta^{\mathrm{mod}\ 2}_{a_{j_1(k)}+a_{j_2(k)}+a_{j_3(k)}+a_{j_4(k)},0} \right) \prod_{i \in C} W(a_{j_1(i)}, a_{j_2(i)}, a_{j_3(i)}, a_{j_4(i)}). \tag{5.1.5}$$

Now,C_B represents the set of all boundary plaquettes, C is the set of all bulk plaquettes, and $j_1(i)$, $j_2(i)$, $j_3(i)$, $j_4(i)$ are links in the plaquette i. The summation $\{a\}$ is taken for all possible configurations of the link variables.

There are two kinds of basic units on the boundary. One of them is the convex hull of a plaquette on the boundary lattice Λ and the closest link to the plaquette in the bulk in $\hat{\Lambda}$ as in Fig. 5.1. We refer to this basic unit as quarter 16-cell. The surface of a quarter 16-cell contains six cells; four of them are tetrahedrons and two of them are

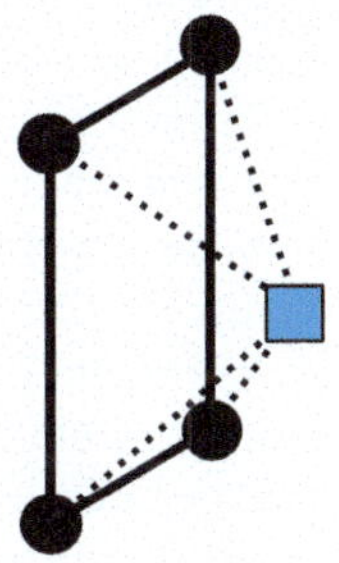

Fig. 5.1 A schematic picture of a quarter 16-cell. The black plaquette represents a plaquette on the boundary and the blue square dot represents a link in the bulk. Reprinted from [1], Copyright 2023, The Author(s) licensed under CC BY 4.0

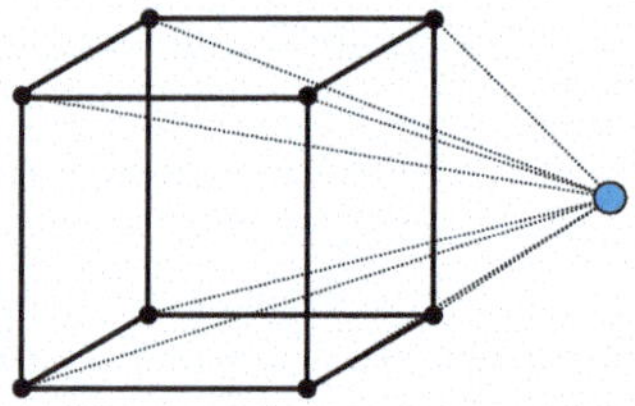

Fig. 5.2 A schematic picture of a cubic cone. The black dots and lines represent sites and links on the boundary, respectively. The blue dot represents a site in the bulk. Reprinted from [1], Copyright 2023, The Author(s) licensed under CC BY 4.0

square pyramids. For example, a quarter 16-cell is the convex hull of the six points $(0, 0, 0, 0)$, $(0, 0, 0, 2)$, $(0, 0, 2, 0)$, $(0, 0, 2, 2)$, $(1, -1, 1, 1)$ and $(1, 1, 1, 1)$.

The other basic unit is the convex hull of a three-dimensional cube on the boundary and a site closest to the cube in the bulk as in Fig. 5.2. We refer to this basic unit as cubic cone. The surface of a cubic cone contains seven cells; six of them are square pyramids and one of them is a cube.

We consider the commutation relation on a cubic cone in Sect. 5.1.2.

5.1.2 Topological Defects Ending on the Boundary

In this subsection, we consider duality defects with edges on the boundary. As we saw in Chap. 4, since duality defects swap active and inactive lattices, the edge of a duality defect on the boundary is expected to be located at the interface between the Neumann N and Dirichlet boundary conditions D or $\widetilde{\mathrm{D}}$. We want to consider the edge of a duality defect that can be topologically deformed on the boundary. Therefore, we impose a "boundary defect commutation relation" between the weights of the elements on the boundary and the edge of the duality defect. We determine their weights in such a way that this condition is satisfied.

In addition to the tetrahedral prism constructed in Chap. 4, there is another kind of building block for KWW duality defects that is adjacent to the boundary. That building block is a doubled square pyramid that has a pair of active and inactive plaquette on the boundary and a pair of active and inactive sites closest to the plaquette in a bulk as shown in Fig. 5.3.

We consider two types of boundary defect commutation relations, one on cubic cone and the other on quarter 16-cell. With this boundary defect commutation relation and the commutation relations considered in Chap. 4, we can construct KWW duality defects that can be smoothly deformed on the boundary. Since the boundary commutation relation on quarter 16-cell does not impose any restriction on the weights of elements on the boundary, we consider the boundary commutation relation on cubic cone in this subsection. The boundary commutation relation on quarter 16-cell is considered in Appendix A.1.

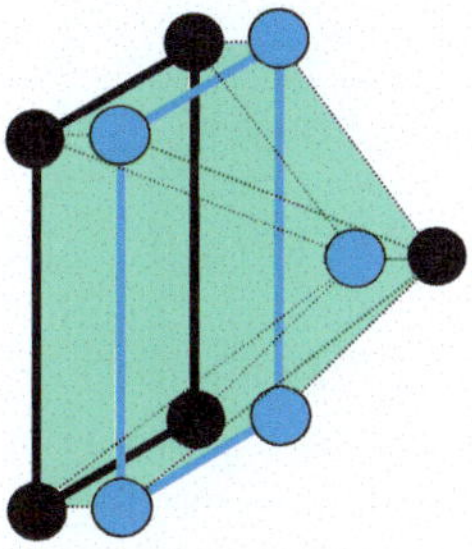

Fig. 5.3 A schematic picture of a building block of KWW duality defects on a boundary. The unit is defined on a doubled square pyramid. Each square pyramid includes a boundary plaquette and a bulk site closest to the plaquette. Reprinted from [1], Copyright 2023, The Author(s) licensed under CC BY 4.0

Consider the boundary defect commutation relation associated with cubic cone. A has six square pyramids on which duality defects can be placed. The boundary defect commutation relation is the equality of one of those configurations with another configurations that does not change the topology. In this case, the defects have the same shape except on cubic cone, which is the one of interest. There are six equations for the boundary commutation relation associated with A, as shown in Fig. 5.4, for the case where the edge exists between the boundary conditions D and N, and between $\widetilde{\mathrm{D}}$ and N, respectively.

The defect commutation relations associated with D and N boundary conditions depicted by Fig. 5.4a–f are given, respectively, by

$$W_{\mathrm{D}}^6 l_{\mathrm{D}}^{12} s_{\mathrm{D}}^8 p_{\mathrm{D}} = W_{\mathrm{D}}^5 l_{\mathrm{D}}^{12} s_{\mathrm{D}}^8 p_{\mathrm{D}}^5, \tag{5.1.6}$$

$$W_{\mathrm{D}}^6 l_{\mathrm{D}}^{12} s_{\mathrm{D}}^8 p_{\mathrm{D}}^2 = W_{\mathrm{D}}^4 l_{\mathrm{D}}^{11} s_{\mathrm{D}}^8 p_{\mathrm{D}}^4, \tag{5.1.7}$$

$$W_{\mathrm{D}}^6 l_{\mathrm{D}}^{12} s_{\mathrm{D}}^8 p_{\mathrm{D}}^3 = W_{\mathrm{D}}^3 l_{\mathrm{D}}^{9} s_{\mathrm{D}}^7 p_{\mathrm{D}}^3, \tag{5.1.8}$$

$$W_{\mathrm{D}}^6 l_{\mathrm{D}}^{12} s_{\mathrm{D}}^8 p_{\mathrm{D}}^3 = W_{\mathrm{D}}^3 l_{\mathrm{D}}^{10} s_{\mathrm{D}}^8 p_{\mathrm{D}}^3, \tag{5.1.9}$$

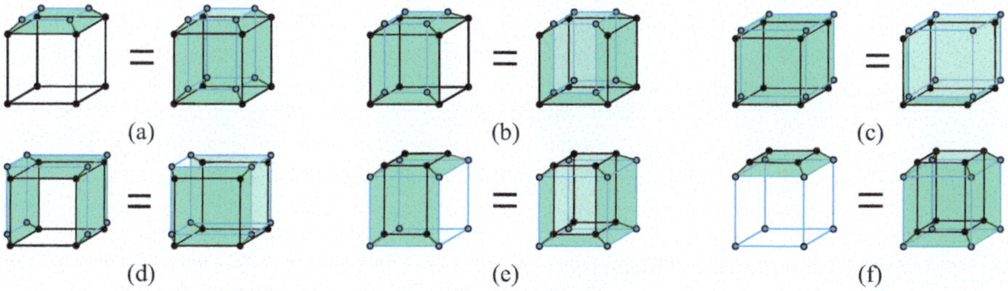

Fig. 5.4 Schematic pictures of boundary defect commutation relations on a cubic cone. Blue lattices represent the N boundary condition. Black lattices represent D or $\widetilde{\mathrm{D}}$ boundary condition. Some of square pyramids are filled by the KWW duality defects which are represented by green surfaces. Reprinted from [1], Copyright 2023, The Author(s) licensed under CC BY 4.0

$$W_{\mathrm{D}}^{2} l_{\mathrm{D}}^{7} s_{\mathrm{D}}^{6} p_{\mathrm{D}}^{2} = W_{\mathrm{D}}^{6} l_{\mathrm{D}}^{12} s_{\mathrm{D}}^{8} p_{\mathrm{D}}^{4}, \tag{5.1.10}$$

$$W_{\mathrm{D}} l_{\mathrm{D}}^{4} s_{\mathrm{D}}^{4} p_{\mathrm{D}} = W_{\mathrm{D}}^{6} l_{\mathrm{D}}^{12} s_{\mathrm{D}}^{8} p_{\mathrm{D}}^{5}. \tag{5.1.11}$$

Note that the Boltzmann weight of the top active plaquette in the right-hand side of Fig. 5.4a does not contribute to the partition function as shown in Eq. (5.1.6), since the inside of this plaquette does not belong to D or $\widetilde{\mathrm{D}}$, but it belongs to N. Equations (5.1.6)–(5.1.11) are not independent. Sorting out these equations, we obtain the following equation, which is equivalent to the above equation:

$$W_{\mathrm{D}} = p_{\mathrm{D}}^{4}, \tag{5.1.12}$$

$$W_{\mathrm{D}}^{3} l_{\mathrm{D}}^{2} = 1, \tag{5.1.13}$$

$$W_{\mathrm{D}}^{3} l_{\mathrm{D}}^{3} s_{\mathrm{D}} = 1. \tag{5.1.14}$$

On the other hand, the defect commutation relations associated with $\widetilde{\mathrm{D}}$ and N boundary conditions depicted by Fig. 5.4a–f are given, respectively, by

$$W_{\widetilde{\mathrm{D}}}^{6} l_{\widetilde{\mathrm{D}}}^{12} s_{\widetilde{\mathrm{D}}}^{8} p_{\widetilde{\mathrm{D}}} = W_{\widetilde{\mathrm{D}}}^{5} l_{\widetilde{\mathrm{D}}}^{12} s_{\widetilde{\mathrm{D}}}^{8} p_{\widetilde{\mathrm{D}}}^{5}, \tag{5.1.15}$$

$$2 W_{\widetilde{\mathrm{D}}}^{6} l_{\widetilde{\mathrm{D}}}^{12} s_{\widetilde{\mathrm{D}}}^{8} p_{\widetilde{\mathrm{D}}}^{2} = W_{\widetilde{\mathrm{D}}}^{4} l_{\widetilde{\mathrm{D}}}^{11} s_{\widetilde{\mathrm{D}}}^{8} p_{\widetilde{\mathrm{D}}}^{4}, \tag{5.1.16}$$

$$2^{3} W_{\widetilde{\mathrm{D}}}^{6} l_{\widetilde{\mathrm{D}}}^{12} s_{\widetilde{\mathrm{D}}}^{8} p_{\widetilde{\mathrm{D}}}^{3} = W_{\widetilde{\mathrm{D}}}^{3} l_{\widetilde{\mathrm{D}}}^{9} s_{\widetilde{\mathrm{D}}}^{7} p_{\widetilde{\mathrm{D}}}^{3}, \tag{5.1.17}$$

$$2^{2} W_{\widetilde{\mathrm{D}}}^{6} l_{\widetilde{\mathrm{D}}}^{12} s_{\widetilde{\mathrm{D}}}^{8} p_{\widetilde{\mathrm{D}}}^{3} = W_{\widetilde{\mathrm{D}}}^{3} l_{\widetilde{\mathrm{D}}}^{10} s_{\widetilde{\mathrm{D}}}^{8} p_{\widetilde{\mathrm{D}}}^{3}, \tag{5.1.18}$$

$$W_{\mathrm{D}}^{2} l_{\mathrm{D}}^{7} s_{\mathrm{D}}^{6} p_{\mathrm{D}}^{2} = 2^{5} W_{\mathrm{D}}^{6} l_{\mathrm{D}}^{12} s_{\mathrm{D}}^{8} p_{\mathrm{D}}^{4}, \tag{5.1.19}$$

$$W_{\mathrm{D}}^{1} l_{\mathrm{D}}^{4} s_{\mathrm{D}}^{4} p_{\mathrm{D}} = 2^{8} W_{\mathrm{D}}^{6} l_{\mathrm{D}}^{12} s_{\mathrm{D}}^{8} p_{\mathrm{D}}^{5}, \tag{5.1.20}$$

where, $\delta^{\mathrm{mod}\ 2}$ and summations of link variables common to both sides are omitted. The numerical coefficients 2^{n} in Eqs. (5.1.16)–(5.1.20) are the numbers of the configurations of the link variables that satisfy the boundary condition. Equations (5.1.15)–(5.1.20) are not independent. Sorting out these equations, we obtain the following equation, which is equivalent to the above equation:

$$W_{\widetilde{\mathrm{D}}} = p_{\widetilde{\mathrm{D}}}^{4}, \tag{5.1.21}$$

$$W_{\widetilde{\mathrm{D}}}^{3} l_{\widetilde{\mathrm{D}}}^{2} = 1, \tag{5.1.22}$$

$$2 W_{\widetilde{\mathrm{D}}}^{3} l_{\widetilde{\mathrm{D}}}^{3} s_{\widetilde{\mathrm{D}}} = 1. \tag{5.1.23}$$

5.1.3 D^3 Expectation Values

We consider a KWW duality defect on D^3 whose edge is S^2 on the boundary, as shown in Fig. 5.5. On the boundary of the spacetime, the inside of this S^2 is Y boundary and the outside of it is X boundary where Y is N and X is D or $\widetilde{\mathrm{D}}$, or vise

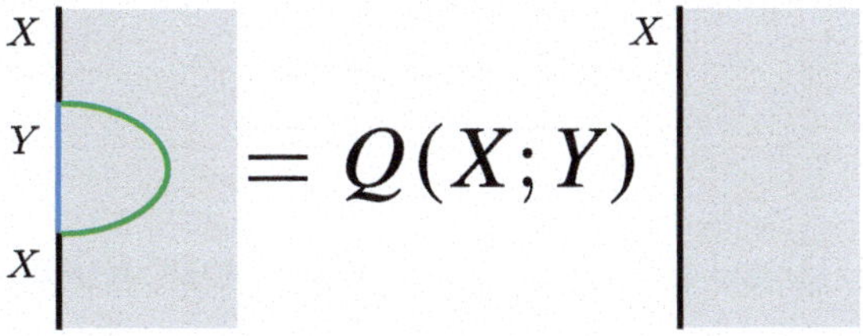

Fig. 5.5 Derivation of the D^3 expectation value. The gray region represents the bulk on which the $\mathbb{Z}_2$ gauge theory lives. The black or blue vertical lines represent boundaries. The green line represents the KWW duality defect on D^3 whose edge is S^2 on the boundary. The black lines represent the X boundary condition. The blue line inside the S^2 represents the Y boundary. When no other operator is contained inside the D^3, we can replace it with D^3 expectation value Q(X; Y). Reprinted from [1], Copyright 2023, The Author(s) licensed under CC BY 4.0

versa by the propriety of duality defect. When this D^3 contains no other operator, it can be replaced by a topological local operator on the boundary. This topological local operator is found to be c-number times the identity operator. This c-number is called "diskv" and is denoted by Q(X; Y). In this subsection, we compute the D^3 Expectation Values. In the following, the D^3 Expectation Values is determined by considering the duality defects placed in on all the square pyramids of a cubic cone, as shown on the left side of Fig. 5.6.

First, we consider Q(N; D) and Q(D; N). The relations of Fig. 5.6 implies the following equations,

$$W_{\mathrm{D}}^6 l_{\mathrm{D}}^{12} s_{\mathrm{D}}^8 p_{\mathrm{D}}^6 = Q(\mathrm{N};\mathrm{D}), \tag{5.1.24}$$

$$W_{\mathrm{D}}^6 l_{\mathrm{D}}^{12} s_{\mathrm{D}}^8 p_{\mathrm{D}}^6 s = Q(\mathrm{D};\mathrm{N}) W_{\mathrm{D}}^6 l_{\mathrm{D}}^{12} s_{\mathrm{D}}^8. \tag{5.1.25}$$

Here, $s = \frac{1}{\sqrt{2}}$. By using Eqs. (5.1.12), (5.1.13), (5.1.14), Q(N; D) and Q(D; N) are expressed in terms of W_{D} as

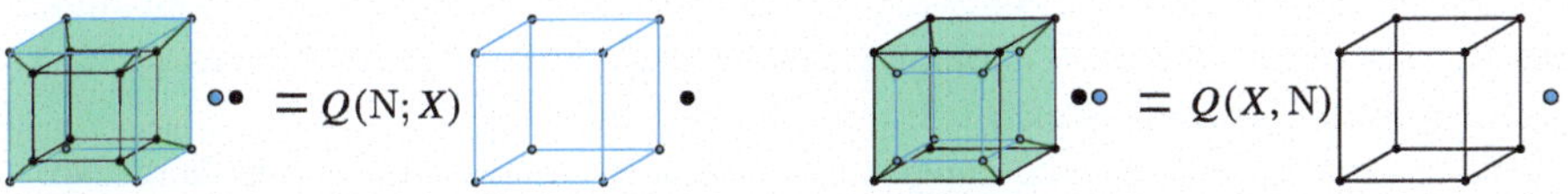

(a) (b)

Fig. 5.6 Configurations used to obtain D^3 Expectation Values. The KWW duality defect connects the N boundary condition with the X boundary condition, where X is D or $\widetilde{\mathrm{D}}$. In the left-hand side, all the square pyramids on the surface of a cubic cone are filled by the KWW duality defect. On the other hand, there is no KWW duality defect in the right-hand side. The roles of X and N are interchanged between (**a**) and (**b**). Reprinted from [1], Copyright 2023, The Author(s) licensed under CC BY 4.0

$$Q(\mathrm{N};\mathrm{D}) = W_{\mathrm{D}}^{\frac{3}{2}}, \tag{5.1.26}$$

$$Q(\mathrm{D};\mathrm{N}) = \frac{1}{\sqrt{2}} W_{\mathrm{D}}^{\frac{3}{2}}. \tag{5.1.27}$$

Consider $Q(\mathrm{N};\widetilde{\mathrm{D}})$ and $Q(\widetilde{\mathrm{D}};\mathrm{N})$. The relations of Fig. 5.6 implies the following equations,

$$2^7 W_{\widetilde{\mathrm{D}}}^6 l_{\widetilde{\mathrm{D}}}^{12} s_{\widetilde{\mathrm{D}}}^8 p_{\widetilde{\mathrm{D}}}^6 = Q(\mathrm{N};\widetilde{\mathrm{D}}), \tag{5.1.28}$$

$$W_{\widetilde{\mathrm{D}}}^6 l_{\widetilde{\mathrm{D}}}^{12} s_{\widetilde{\mathrm{D}}}^8 p_{\widetilde{\mathrm{D}}}^6 s = Q(\widetilde{\mathrm{D}};\mathrm{N})\, W_{\widetilde{\mathrm{D}}}^6 l_{\widetilde{\mathrm{D}}}^{12} s_{\widetilde{\mathrm{D}}}^8. \tag{5.1.29}$$

Here, the coefficient 2^7 on the left side of Eq. (5.1.28) is the number of possible configurations of the boundary link variables that satisfy the boundary condition. By using the relations (5.1.21), (5.1.22), (5.1.23), the D^3 Expectation Values $Q(\mathrm{N};\widetilde{\mathrm{D}})$, $Q(\widetilde{\mathrm{D}};\mathrm{N})$ are expressed in terms of $W_{\widetilde{\mathrm{D}}}$ as

$$Q(\mathrm{N};\widetilde{\mathrm{D}}) = \frac{1}{2} W_{\widetilde{\mathrm{D}}}^{\frac{3}{2}}, \tag{5.1.30}$$

$$Q(\widetilde{\mathrm{D}};\mathrm{N}) = \frac{1}{\sqrt{2}} W_{\widetilde{\mathrm{D}}}^{\frac{3}{2}}. \tag{5.1.31}$$

The above calculations show that the duality defect with D^3 topology with no other operator inside can be replaced by the D^3 expectation value times the identity operator. We represent the D^3 Expectation Values in terms of the Boltzmann weight on the boundary.

5.1.4 Relations Between G-Functions

In this subsection, we find the ratio of g-functions with different boundary conditions. In the AMF approach, it is not easy to obtain a fusion rule for the boundary and duality defects. Therefore, the ratio is obtained using the D^3 Expectation Values $Q(\mathrm{X};\mathrm{Y})$ obtained in the previous Sect. 5.1.3.

We consider the $\mathbb{Z}_2$ lattice gauge theory on a 4-dimensional hemisphere. Next, we place a duality defect on D^3 ending on S^2 on the boundary of the 4-dimensional hemisphere. The boundary conditions change from D to N or from $\widetilde{\mathrm{D}}$ to N at the edge of this KWW duality defect. We denote the partition function with this defect by $\langle V \rangle$. This defect can be replaced by the D^3 expectation value. In particular, since this spacetime is a 4-dimensional hemisphere, as shown in the Fig. 5.7, this identity can be used in two different ways. As a result, we obtain the relation:

$$g_X Q(X;Y) = \langle V \rangle = g_Y Q(Y;X), \tag{5.1.32}$$

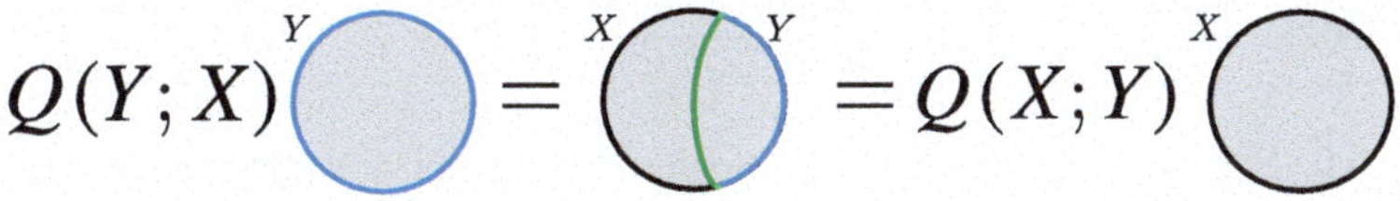

Fig. 5.7 The derivation of the relation between g-functions. The green line represents the KWW duality defect on D^3 ending on S^2 on the boundary of the four-dimensional hemisphere on which the $\mathbb{Z}_2$ lattice gauge theory lives. This KWW duality defect connects the boundary conditions X and Y. The black boundary represents the boundary condition X, and the blue boundary represents the boundary condition Y. We use the identity in Fig. 5.5 in two different ways: the left-hand side and the right-hand side. Reprinted from [1], Copyright 2023, The Author(s) licensed under CC BY 4.0

where X and Y denote the two boundary conditions connected by the duality defect. From Eq. (5.1.32) for $(X, Y) = (\mathrm{D}, \mathrm{N})$ and $(X, Y) = (\widetilde{\mathrm{D}}, \mathrm{N})$, we obtain the following relations,

$$g_{\mathrm{D}} Q(\mathrm{D}; \mathrm{N}) = g_{\mathrm{N}} Q(\mathrm{N}; \mathrm{D}), \tag{5.1.33}$$

$$g_{\widetilde{\mathrm{D}}} Q(\widetilde{\mathrm{D}}; \mathrm{N}) = g_{\mathrm{N}} Q(\mathrm{N}; \widetilde{\mathrm{D}}). \tag{5.1.34}$$

Here, g_{D}, $g_{\widetilde{\mathrm{D}}}$, and g_{N} denote the g-functions of D, $\widetilde{\mathrm{D}}$, and N, respectively. From these equations and the expressions of the D^3 Expectation Values (5.1.26), (5.1.27), (5.1.30), (5.1.31), we obtain the relations between g-functions:

$$\frac{1}{2} g_{\mathrm{D}} = \frac{1}{\sqrt{2}} g_{\mathrm{N}} = g_{\widetilde{\mathrm{D}}}. \tag{5.1.35}$$

According to the 4-dimensional g-theorem, the g-function is monotonically increasing along the boundary renormalization group flow. Therefore, the relation (5.1.35) implies that the boundary renormalization group flows from $\widetilde{\mathrm{D}}$ to N and from N to D are prohibited.

It is not easy to obtain the fusion rules for the boundary and duality defects in the AMF approach. But we obtain them indirectly from Eq. (5.1.35) and the S^3 expectation value of the duality defect $\frac{1}{\sqrt{2}}$:

$$\mathrm{D} \times K = \mathrm{N}, \quad \mathrm{N} \times K = \widetilde{\mathrm{D}}, \tag{5.1.36}$$

where K is the KWW duality defect. This is also consistent with the bulk fusion rule $K \times K = C$, where C is the codimension one condensation defect of the $\mathbb{Z}_2$ one-form symmetry [2, 3]; the fusion rule $\mathrm{D} \times C = \widetilde{\mathrm{D}}$ derived from this bulk fusion rule and Eq. (5.1.36) agrees with the definitions of D and $\widetilde{\mathrm{D}}$.

References

1. Koide, M., Nagoya, Y., Yamaguchi, S.: Noninvertible symmetries and boundaries in four dimensions. Phys. Rev. D **108**(6), 065009 (2023). arXiv: 2304.01550 [hep-th]
2. Roumpedakis, K., Seifnashri, S., Shao, S.-H.: Higher gauging and non-invertible condensation defects (2022). arXiv: 2204.02407 [hep-th]
3. Choi, Y., et al.: Non-invertible condensation, duality, and triality defects in 3+1 dimensions (2022). arXiv: 2204.09025 [hep-th]

Chapter 6
Conclusion and Discussion

Here we summarize this Ph.D. thesis. In Chap. 2, we reviewed generalized symmetries. Traditionally, symmetry has been considered as an invariance of the action associated with a transformation. Then, when there exists a continuous symmetry, we can define a conservative current from Noether's theorem and use the current to derive the WT-identity. The WT-identity then implies locally global symmetry transformations for the field. In order to consider such local global transformations for symmetries including discrete symmetries, we define a symmetry defect from the variation of the action of a global transformation on a region of spacetime. Symmetry defects are topological because the region over which the global transformation is performed can be changed by replacing the labels of the integral variables. The symmetry defects have codimension-1 and are invertible. In generalized symmetry, we consider general topological defects as symmetries. As concrete examples, this chapter introduces, in particular, higher-form symmetries whose codimension is not 1 and non-invertible symmetries consisting of non-invertible defects.

In Chap. 3 we have considered the 2-dimensional Ising model. First, we have proved KW duality, which implies equivalence between the Ising model in the 2-dimensional, high-temperature phase (low-temperature phase) and the Ising model with $\mathbb{Z}_2$ symmetry gauged in the low-temperature phase (high-temperature phase). Then, following the Aasen-Mong-Fendley approach, we introduced a dual lattice and a KW duality defect placed between the KW dual theories. We imposed a commutation relation on this defect and solved it to construct a topological defect. By calculating the expectation value of a topological defect placed on S^1, we have confirmed that the topological defect is non-invertible. We have also constructed symmetry defects corresponding to 0-form $\mathbb{Z}_2$ symmetry and junctions between them and non-invertible defects, and considered the relation between defects with different topologies.

In Chap. 4, we introduced one of the results of our study, a non-invertible defect in the 4-dimensional $\mathbb{Z}_2$ lattice gauge theory. This model has 1-form $\mathbb{Z}_2$ symmetry and KWW duality. By applying the Aasen-Mong-Fendley approach, we have introduced

M. Koide, *Non-invertible Symmetry in 4-Dimensional Z2 Lattice Gauge Theory*,
Springer Theses, https://doi.org/10.1007/978-981-96-2272-6_6

a dual lattice and KWW defects placed between the KWW dual theories. Specifically, we focus on a single 16-cell consisting of a pair of active and inactive plaquettes on our setup, which consists of an active and inactive lattice. We defined a building block of KWW duality defects on the tetrahedral prism, which is a doubling of the tetrahedron consisting of pairs of active and inactive links on the surface of the 16-cell. Then, topological defects were constructed by imposing and solving a commutation relation on the 16-cell to that defects, which implies the topological nature of the defects. We also placed the defects on all the tetrahedrons existing on the surface of the 16-cell and calculated the expectation value of the KWW duality defect with S^3 topology. By comparing the expectation values with the weights of the empty 16-cell, we proved that the KWW duality defects are non-invertible. We also constructed a 1-form $\mathbb{Z}_2$ global symmetry defect and a topological junction connecting the non-invertible duality defect and the 1-form symmetry defect. We also have obtained relationships between defects with different topologies and used these relationships to calculate the expectation values of non-invertible duality defects in several configurations.

In Chap. 5, we consider gauge theories with boundary. We defined one type of Neumann boundary condition on a boundary made of inactive lattices and two types of Dirichlet boundary conditions on a boundary made of active lattices. The KWW duality was defined to have an edge on the boundary between the Neumann and Dirichlet boundary conditions due to a defect that swaps the active and inactive lattices. We also defined the edge of the defect to be topological on the boundary. We then considered a D^3 KWW duality defect with a S^2 edge on the boundary. Then, if no operator exists in the region bounded by this defect on D^3 and the D^3 boundary, we could replace it with the c-number, the D^3 expectation value. After the replacement, the boundary conditions of the theory became the boundary conditions defined outside of S^2 before the defect was replaced. This D^3 expectation value depended on the boundary conditions inside the outside of the edge of the defect before the deformation. We then considered a D^3 defect in a 4-dimensional hemisphere with S^2 edges at the boundaries. We then replaced the defects with D^3 expectation values in two ways to obtain the ratios of the hemisphere partition function for the Neumann boundary condition to the hemisphere partition function for the Dirichlet boundary condition. Then, from the results and the g-theorem, we obtained a restriction on the renormalization group flow.

Here we consider another solution to the duality defect of the Ising model in two dimensions. We used the solutions of $D(a, \tilde{a}) = 1/\sqrt{2}(-1)^{a\tilde{a}}$ in Sect. 3.3, and these solutions with different signs are also solutions of the defect commutation relation. One specific example is,

$$D(a, \tilde{a}) = -\frac{1}{\sqrt{2}}(-1)^{a\tilde{a}}. \tag{6.1}$$

These solutions are indistinguishable if the entire defect consists of an even number of building blocks, but if it consists of an odd number of building blocks, they differ only in the sign of the whole. For example, the odd number of defects are composed on the projective plane.

The following solutions are also alternative solutions,

$$D(a,\tilde{a}) = \frac{1}{\sqrt{2}}(-1)^{(1-a)(1-\tilde{a})}, \tag{6.2}$$

$$D(a,\tilde{a}) = \frac{1}{\sqrt{2}}(-1)^{(1-a)\tilde{a}}, \qquad D(a,\tilde{a}) = \frac{1}{\sqrt{2}}(-1)^{a(1-\tilde{a})}. \tag{6.3}$$

The solution (6.2) satisfies $D(a,\tilde{a}) = D(\tilde{a},a)$. While the solutions (6.3) do not satisfy $D(a,\tilde{a}) = D(\tilde{a},a)$ These solutions are considered to be identical if the defect does not have an edge. For example, consider a another solution $\tilde{D}(a,\tilde{a}) = 1/\sqrt{2}(-1)^{(1-a)\tilde{a}}$. To distinguish it from the original solution $D(a,\tilde{a}) = 1/\sqrt{2}(-1)^{a\tilde{a}}$, the another solution is marked with the tilde. Assuming that the defect does not have an edge, there is another building block $\tilde{D}(b,\tilde{a})$ that contains the active site $\tilde{a}$. This makes it indistinguishable from the original solution, as in the following equation.

$$\tilde{D}(a,\tilde{a})\tilde{D}(b,\tilde{a}) = \frac{1}{2}(-1)^{(1-a)\tilde{a}}(-1)^{(1-b)\tilde{a}} = \frac{1}{2}(-1)^{a\tilde{a}}(-1)^{b\tilde{a}} = D(a,\tilde{a})D(b,\tilde{a}). \tag{6.4}$$

Next, we discuss the prospects for this study. One of the directions is to apply the AMF approach to discover more non-invertible symmetries. In this thesis, applying the AMF approach, we construct topological defects corresponding to the KWW duality transformation. Another approach to construct KWW duality defects is half space gauging. In half-space gauging, we consider gauging only for partial regions of spacetime, as in the global transformation on partial regions performed in Chap. 2. Since the gauging is performed by adding up the symmetry defects in all topologically different configurations, this region can be topologically deformed. If the theory is self-dual to the gauging, the gauged region is equal to the original theory, so the effects of the deformation remain as non-invertible topological operators at the boundaries of the region. Thus, in half-space gauging, self duality is a necessary condition for constructing topological defects. On the other hand, in the AMF approach, the equality of two theories separated by a defect is set first. A topological nature is then satisfied by imposing a defect commutation relation. Therefore, self-duality is not a necessary condition for the construction of non-invertible topological defects. Therefore, we consider it to be a powerful tool for finding non-invertible symmetries based on properties other than self-duality associated with gauging. In fact, by the AMF approach, such defects are constructed in two and three dimensions [1, 2].

References

1. Aasen, D., Fendley, P., Mong, R.S.K.: Topological defects on the lattice: dualities and degeneracies (2020). arXiv: 2008.08598 [cond-mat.stat-mech]
2. Inamura, K., Ohmori, K.: Fusion surface models: 2+1d lattice models from fusion 2-categories (2023). arXiv: 2305.05774 [cond-mat.str-el]

Appendix

A.1 Commutation Relations on Quarter 16-Cell

We examine boundary defect commutation relations on a quarter 16-cell. A quarter 16-cell contains four tetrahedrons and two square pyramids as three-dimensional elements. We focus on one quarter 16-cell and consider placing building blocks of the duality defect on three-dimensional elements on it. There are four boundary defect commutation relations of the KWW duality defect connecting boundary conditions D and N or $\widetilde{\mathrm{D}}$ and N as shown in Fig. A.1.

All of these boundary defect commutation relations are satisfied for arbitrary values of the weights of the boundary elements if we use the bulk weights obtained in Sect. 4.2. For example, the boundary defect commutation relations of Fig. A.1c connecting $\widetilde{\mathrm{D}}$ and N is the following equation.

$$\sum_{a_1,a_2,a_3,a_4} W_{\widetilde{\mathrm{D}}}\delta^{\mathrm{mod}\ 2}_{a_1+a_2+a_3+a_4,0} p_{\widetilde{\mathrm{D}}} s^4_{\widetilde{\mathrm{D}}} l^4_{\widetilde{\mathrm{D}}} s$$
$$= \sum_{a_1,a_2,a_3,a_4,a_5} W_{\widetilde{\mathrm{D}}}\delta^{\mathrm{mod}\ 2}_{a_1+a_2+a_3+a_4,0} p_{\widetilde{\mathrm{D}}} s^4_{\widetilde{\mathrm{D}}} l^4_{\widetilde{\mathrm{D}}} s^2 l^1 D(a_5,a_1) D(a_5,a_2) D(a_5,a_3) D(a_5,a_4). \tag{A.1.1}$$

Here $a_1,\ a_2,\ a_3,\ a_4$ are boundary link variables, a_5 is a bulk link variable and $l = s = \frac{1}{\sqrt{2}}$ are weights in the bulk. This equation is an identity with respect to $W_{\widetilde{\mathrm{D}}}, p_{\widetilde{\mathrm{D}}}, s_{\widetilde{\mathrm{D}}}, l_{\widetilde{\mathrm{D}}}$.

M. Koide, *Non-invertible Symmetry in 4-Dimensional Z2 Lattice Gauge Theory*, Springer Theses, https://doi.org/10.1007/978-981-96-2272-6

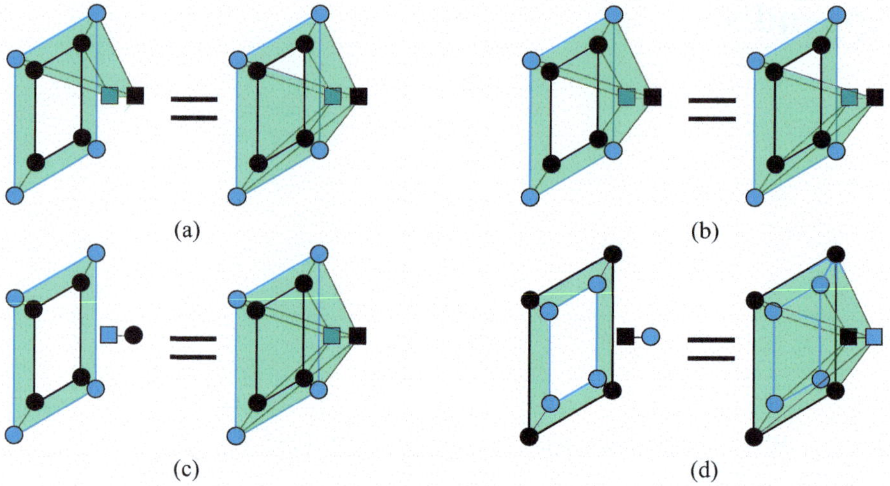

Fig. A.1 Boundary defect commutation relations on a quarter 16-cell. Black plaquettes represent plaquettes on the boundary with D or $\widetilde{\mathrm{D}}$ boundary condition and blue plaquettes represent plaquettes on the boundary with N boundary condition. Black square dots and blue square dots represent active links and inactive links in the bulk, respectively. Black circular dots and blue circular dots represent active sites and inactive sites, respectively. green surfaces represent the KWW duality defects. Some of four tetrahedrons are filled by the KWW duality defect. One square pyramid out of two is filled with KWW defects, which are omitted in the figure. Reprinted from [1], Copyright 2023, The Author(s) licensed under CC BY 4.0

Reference

1. Koide, M., Nagoya, Y., Yamaguchi, S.: Noninvertible symmetries and boundaries in four dimensions. Phys. Rev. D **108**(6) 065009 (2023). arXiv: 2304.01550 [hep-th]

The manufacturer's authorised representative in the EU is Springer Nature Customer Service Centre GmbH, Europaplatz 3, 69115 Heidelberg, Germany. If you have any concerns regarding our products, please contact ProductSafety@springernature.com

Printed and bound by CPI Group (UK) Ltd, Croydon, CR0 4YY
01/10/2026
02225945-0004